Quality Assurance Program on Stereotactic Radiosurgery

AF906138

Springer
Berlin
Heidelberg
New York
Barcelona
Budapest
Hong Kong
London
Milan
Paris
Tokyo

Günther H. Hartmann

Quality Assurance Program on Stereotactic Radiosurgery

Report from a Quality Assurance Task Group

Contributing Authors:
Wendell Lutz, Jürgen Arndt, Igor Ermakov, Ervin B. Podgorsak,
Lothar Schad, Christopher Serago, Stanislav M. Vatnitsky

With 37 Figures

Springer

Günther H. Hartmann, Ph. D. (Chairman)
Department of Radiology
German Cancer Research Center
Im Neuenheimer Feld 280
69120 Heidelberg
Germany

ISBN-13: 978-3-540-59199-3 e-ISBN-13: 978-3-642-79718-7
DOI: 10.1007/978-3-642-79718-7

Library of Congress Cataloging-in-Publication Data. Hartmann, Günther H.: Quality assurance program on stereotactic radiosurgery: report from a quality assurance task group / Günther H. Hartmann. Contributing authors: Wendell Lutz... - Berlin; Heidelberg; New York; Barcelona; Budapest; Hong Kong; London; Mailand; Paris; Tokyo: Springer, 1995
33 DBN 94.459123.X 95.06.08 6785 ko

This work is subject to copyright. All rights are reserved, whether the whole or part of the material is concerned, specifically the rights of translation, reprinting, reuse of illustrations, recitation, broadcasting, reproduction on microfilm or in any other way, and storage in data banks. Duplication of this publication or parts there of is permitted only under the provisions of the German Copyright Law of September 9, 1965, in its current version, and permisstion for use must always be obtained from Springer-Verlag. Violations are liable for prosecution under the German Copyright Law.

© Springer-Verlag Berlin Heidelberg 1995

The use of general descriptive names, registered names, trademarks, etc. in this publication does not imply, even in the absence of a specific statement, that such names are exempt from the relevant protective laws and regulations and therefore free for general use.

Product liability: The publishers cannot guarantee the accuracy of any information about the application of operative techniques and medications contained in this book. In every individual case the user must check such information by consulting the relevant literature.

SPIN: 10133994 21/3135 - 5 4 3 2 1 0 - Printed on acid-free paper

Contributors

Jürgen Arndt, M. Sc.
Elekta Instrument AB, POB 7593, S-103 93 Stockholm, Sweden

Igor Ermakov, Ph. D.
Central Scientific Research Institute for Roentgenology and Radiology, Pesochnij-2, ul. Leningradskaja, 70/4, 188646, St Petersburg, Russia

Günther H. Hartmann, Ph. D. (Chairman)
Department of Radiology, German Cancer Research Center, Im Neuenheimer Feld 280, 69210 Heidelberg, Germany

Wendell Lutz, Ph. D.
Radiation Oncology, University of Arizona, 1501 North Campbell Avenue, Tucson, AZ 85724, USA

Ervin B. Podgorsak, Ph. D.
Department of Radiation Oncology, McGill University, 1650, avenue des Cedres, Montreal, Quebec H3G 1A4, Canada

Christopher Serago, Ph. D.
Massachusetts General Hospital and Harvard Medical School, Department of Radiation Medicine, Boston, MA 02114, USA

Stanislav M. Vatnitsky, Ph. D.
Loma Linda University Medical Center, Department of Radiation Medicine, Loma Linda, CA 92354, USA

Preface

At the end of an international workshop on "Proton and Narrow Photon Beam Therapy" which took place in Oulu, Finland in 1989, a "Quality Assurance Task Group" was established. The members of this group were: Jürgen N. Arndt, Bernhard Bauer-Kirpes, Giorgio Chierego, Igor Ermakov, Günther H. Hartmann, Wendell Lutz, Christina Marchetti, Mark H. Phillips, Ervin B. Podgorsak, Christopher Serago, Stanislav M. Vatnitsky, and H. Treuer. All group members are medical physicists, actively involved in establishing stereotactic radiosurgery and in making it an accepted therapeutic technique. The aim was to share the essence of their experience with other interested parties with a view to developing and applying different physical and technical solutions for radiosurgery.

The group has discussed and established recommendations and guidelines for quality assurance programs, especially with regard to those procedures which are physical or technical in character and specific to radiosurgery. The results are now summarized in this document. The recommendations have no official status, but are rather intended as suggestions to colleagues who are already involved in or are in the process of entering the field of radiosurgery.

This document identifies of those parameters which are essential for radiosurgery, i.e., the selective and reproducible dose delivery to small volumes of tissue. Although mainly conceived for stereotactic radiosurgery, some of these parameters may also be applicable to fractionated stereotactic radiotherapy. As far as possible, specifications of tolerances and also practical solutions to their control and verification are given. Some of the issues are applicable regardless of technique and are discussed in general terms. Others are not and are therefore dealt with more specifically.

The first chapter of this document contains specific definitions required to provide a consistent treatment throughout the report. The second chapter is devoted to general principles, with the focus on quality assurance. In chapter three, equipment specific quality assurance tests are described which should precede clinical application of the relevant radiosurgical technique.

Although this report is focused on quality assurance for the technical aspects of radiosurgery it must be emphasized that there are a number of other factors of great importance, which must not be forgotten. An example of such a factor, which is generally not initially identified, is the complexity of the procedures and the variety of disciplines involved. Radiosurgery, more than many other medical procedures, requires a team of specialists who are highly skilled in their own professional fields but who also have a common understanding of the other disciplines involved.

Although many patients will and have benefited from radiosurgery, the procedure is potentially dangerous. The word surgery is misleading in the sense that it is associated with a procedure that has consequences which generally are predictable, limited in time and extent. Our knowledge regarding the consequences of radiation absorbed in tissue is mainly based on empirically gained information. Thus, in the first place deviations between the planned and delivered dose may become manifest years after the delivery of that irradiation in the form of irreversible side effects. Moreover, changes in irradiation schedules may have similar effects. Experience from radiotherapy tells that as long as 20 years may pass before side effects caused by radiation occur. Therefore, our awareness and monitoring of side effects must be on a longer time scale.

Radiosurgery is today an established and accepted method of treatment, but as in all dynamic fields, there is debate concerning basic questions. For example, there is still some controversy as to what are the best indications for radiosurgical treatment. It remains to be determined what is the ideal dose distribution within different lesions and which dose is optimal or tolerable under different conditions. Future developments will be based on current scientific research.

We wish to gratefully acknowledge the financial support for this project from the following companies: Elekta Instrument AB, Stockholm, Sweden; Leibinger GmbH, Freiburg, Germany; Varian, Medical & Industrial Group, Palo Alto, CA 94303, USA.

G. H. Hartmann, Heidelberg, Spring 1995

Contents

1 Introduction

1.1 Purpose of the Report

The purpose of this report is to describe quality assurance (QA) programs with respect to the physical aspects of stereotactic radiosurgery through identifying the sources of errors and suggesting quality assurance tests that should be performed in order for a stereotactic system to be used clinically. To this end, quality assurance programs are needed at the time of implementation of the radiosurgical technique, periodically thereafter, and before and during each treatment.

Stereotactic radiosurgery is performed with a range of techniques and radiation types. This report will cover quality assurance (QA) programs only for the two main techniques which are based on a gamma unit ("Gamma Knife") or on a linac facility. No attempt was made to describe the various methods that have been developed to perform stereotactic radiosurgery in detail. They can be found in review articles or textbooks [39,42,61]. The quality assurance programs and/or tests are presented, when possible, independently from particular radiosurgical techniques. However, some tests described will only be applicable to gamma units or linac-based facilities; others are applicable to both; particle beam issues are not discussed.

Stereotactic radiosurgery comprises a chain of single procedures which is also used to structure this report. The following main procedures which are common to all radiosurgical techniques are addressed:

1. Stereotactic localization and immobilization
2. Dosimetry and treatment planning
3. Treatment delivery

Quality assurance tests will be recommended for each of these procedures on three levels. The first level comprises tests used for the implementation and commissioning of a stereotactic system. The tests of the second level should be repeated periodically to ensure continued reliable performance. The third level of tests comprises procedures to be used before and during each particular treatment.

1.2 Definition of Key Terms

In the context of stereotactic radiosurgery some terms such as accuracy and precision, isocenter, target point and volume, are frequently used. A specific definition of each is required to ensure a consistent treatment throughout the report. The definitions, as used in this report, are given below.

1.2.1 Accuracy and Precision

The concepts of accuracy and precision are of particular importance in stereotactic radiosurgery. It needs to be emphasized, however, that any reasonable statement on accuracy or precision should be adequately analyzed and reported only by statistical means. Therefore, repeated measurements of a particular parameter are always required.

Accuracy denotes the closeness of agreement between the "true" value of a test variable and its measurement. Therefore, in the absence of the knowledge of the "true" value, the accuracy of any measurement cannot be known. On the other hand, if the true value is known, the best estimate of accuracy, or inaccuracy, is given by the difference between the mean value of repeated measurements of a variable and its true value. Inaccuracy directly refers to the systematic errors which may occur at a specific radiosurgery facility or procedure. In principle some but not all systematic inaccuracies can be eliminated [23].

Precision refers to the reproducibility of a measurement. The degree of precision may be expressed by the standard deviation of the test variable and provides a direct measure of the consistency of a procedure.

1.2.2 The Isocenter Concept

Common to all radiosurgical irradiation techniques is the use of cross firing or convergent beam irradiation techniques. Under ideal conditions all beam central axes intersect at a common point in space. At linear accelerator facilities, a prerequisite for this is that the gantry and couch rotation axes meet at one point, which is referred to as the machine isocenter. In practice all these axes do not intersect at a common point but only come closest together within a "best compromise" sphere. The reference setup point is defined as point located within this sphere and represents the location at which the target point of a patient is placed for treatment. Its exact location within this sphere is determined operationally and depends on the radiosurgical technique used. A similar "best compromise" sphere can be defined for the gamma unit as the intercept of the central axes of all Co-60 beams. Its center is referred to as the "unit center point" (UCP).

1.2.3 Target Point

The target point is that point (or points) within the treatment volume which is placed at the reference setup point for treatment. It is obtained as a result of the treatment planning process simulating the irradiation setup. Therefore, it should not be determined from the diagnostic information alone.

1.2.4 Volumes

Conceptual definitions on volumes involved in radiotherapy have been recently given in ICRU 50 [26]. The "clinical target volume" is the actual volume of the diseased tissue as defined by the clinician. The "planning target volume" is defined as the clinical target volume plus a margin that reflects the spatial uncertainties in localization and dose delivery techniques. "Treated volume" is defined by the volume enclosed by the prescription isodose surface, selected and specified by the radiation oncologist as being appropriate to achieve the purpose of treatment.

2 Quality Assurance Aspects and Requirements

2.1 Quality Assurance Aspects of Stereotactic Target Localization

2.1.1 Background

The concept behind using a stereotactic technique is twofold: (1) to introduce a three-dimensional coordinate system affixed to the patient, and (2) to accurately locate and access any desired region or point within the patient's brain (or other regions). This so-called stereotactic coordinate system may be a Cartesian, cylindrical, or spherical coordinate system. The stereotactic guiding device system is normally based on the use of a specific stereotactic head frame. Stereotactic devices are currently available for a wide variety of intracranial neurosurgical indications ranging from biopsy to tumor resection and interstitial brachytherapy. For use in radiosurgery, such devices have been adapted to be compatible with the radiosurgical technique used [9,20-22,25,29,31,35,36,38,53,57].
Generally, the stereotactic frame is a rigid device in the form of a hollow cube or cylinder, which is fastened with pins to the patient's head. Thus it provides: (1) a unique and fixed relationship between the frame and the brain, (2) the means by which this reference frame can be applied to medical images of the brain, (3) accurate and precise positioning for as many diagnostic and treatment devices and procedures as necessary. Any arbitrary point within the cranium can be determined with respect to the origin of the stereotactic frame. Subsequently, assuming that the frame remains rigidly attached to

the cranium, the same points may be accessed either invasively (stereotactic biopsy or surgery) or non-invasively with radiation beams (radiosurgery). However, the frame may slip, particularly with relocatable frames. Therefore, the assumption that a frame does not move during the various steps of the radiosurgical procedure cannot be taken for granted and the user is advised to test this assumption.

The stereotactic target localization procedure provides the patient data for treatment planning, i.e., anatomical location and shape of the clinical and planning target volume with one or more imaging modalities. For this purpose, two requirements are important: (1) the head frame must not interfere with the ability of the imaging system to localize the target, (2) an accurate method for converting the imaging coordinates into the stereotactic coordinates of the frame must be available. In particular, the method must be able to account for tilt angles between image coordinates and stereotactic coordinates.

The accurate determination of stereotactic coordinates from imaging coordinates may depend upon several factors or processes which should be carefully reviewed. For example, algorithms imbedded in the planning software may assume that CT or MRI imaging planes are parallel to the head ring. Manufacturing defects or mechanical stresses may cause a distortion of the localization devices which could introduce errors. Another potential localization error would result from inadequately mounting the localization device to the head ring. Therefore, tests should be available to evaluate the correct transformation of imagine coordinates to stereotactic coordinates. Suggestions for such tests will be provided in the third chapter.

Several manufacturers world-wide now offer stereotactic frames that are compatible with modern imaging equipment (CT, MRI) and can be adapted for radiosurgery. Compatibility with CT implies that the frame is made of low-atomic-number materials, such as plastics or aluminum, to minimize the interference of the frame with the diagnostic CT X - ray beam. An MRI-compatible frame is made of non-ferrous materials and designed in such a way that closed-current loops within the frame structure are avoided during the MR imaging. However, this does not guarantee distortion-free MR images for the purpose of target localization. For detailed descriptions of stereotactic localization techniques the reader is referred to the literature or manuals provided by the manufacturers.

The degree of accuracy and precision found with different imaging modalities using specific methods for target localization varies from a few tenths of a millimeter up to several millimeters. Therefore, each specific localization method requires an initial acceptability test before use.

2.1.2 Key Parameters and Requirements

Stereotactic localization procedures require high spatial accuracy and precision. The following discussion describes an quantitative approach to determining spatial accuracy and precision. To this end three key parameters will be introduced.

The approach consists of the measurement of the difference between the "true" coordinates (mechanically measured) and the calculated coordinates of test points fixed within the CT, MRI or plain film localization devices as used clinically. Let the position of this "true" point in space be expressed by the stereotactic coordinates x_t, y_t, and z_t. The coordinates of this point are measured N times by applying the localizing procedure of interest. At the *i* th measurement the coordinates x_i, y_i, and z_i are obtained. Then the following three key parameters are recommended for reporting the spatial accuracy and precision of a stereotactic localization procedure.

2.1.2.1 First Parameter

The length, $\bar{r}$, of the mean of all vectorial displacements between the true and the measured points in space.

This parameter is particularly useful for detecting systematic uncertainties in a procedure. It is calculated as:

$$\bar{r} = \frac{1}{N}\sqrt{\left(\sum_{i=1}^{N}(x_t - x_i)\right)^2 + \left(\sum_{i=1}^{N}(y_t - y_i)\right)^2 + \left(\sum_{i=1}^{N}(z_t - z_i)\right)^2}$$

If $\bar{r}$ equals zero, then only random errors exist in a particular localization procedure.

2.1.2.2 Second Parameter

The mean, $\bar{r}_i$, of all displacement errors between the true point and the measured points.

This parameter characterizes the accuracy of a single point measurement, i.e., $\bar{r}_i$ is the best estimate of a likely error expected from one single localization. It is calculated as the mean of all displacements:

$$\bar{r}_i = \frac{\sum_{i=1}^{N} r_i}{N}$$

where r_i is the length of the error vector at the i th measurement

$$r_i = \sqrt{(x_t - x_i)^2 + (y_t - y_i)^2 + (z_t - z_i)^2}$$

2.1.2.3 Third Parameter

The standard deviation, s_r, of the second parameter, the mean of all displacement errors.

This parameter characterizes the precision of the localization procedure. It is calculated from a sample of N single measurements as:

$$s_r = \sqrt{\frac{\sum_{i=1}^{N} (\bar{r}_i - r_i)^2}{N - 1}}$$

These three key parameters may be applied not only to the localization procedures, but also to overall measurements of spatial accuracy and precision including localization and treatment. In this case, the second and the third parameter, $\bar{r}_i$ and s_r, are very important with respect to the delineation of the planning target volume, since a combination of them provides a measurement of the margin necessary between the clinical target volume and the planning target volume. For example, if the prescriptive isodose surface just matches the planning target volume, the separation between the clinical target volume and the planning target volume should be approximately $\bar{r}_i + 2 \cdot s_r$. This will ensure that 95% of all such treated patients have clinical target volumes that lie entirely within the prescriptive isodose surface. Figures 2.1.a and 2.1.b show four cases with uncertainties from different sources for which the application of the three key parameters is exemplified. The results obtained for the key parameters are shown in Table 2.1.

Table 2.1 Results obtained for the key parameters

Key Parameter	Example I	II	III	IV
$\bar{r}$	0	0	0.5	0.5
$\bar{r}_i$	0.3	0.8	0.6	0.9
s_r	0.14	0.34	0.18	0.37

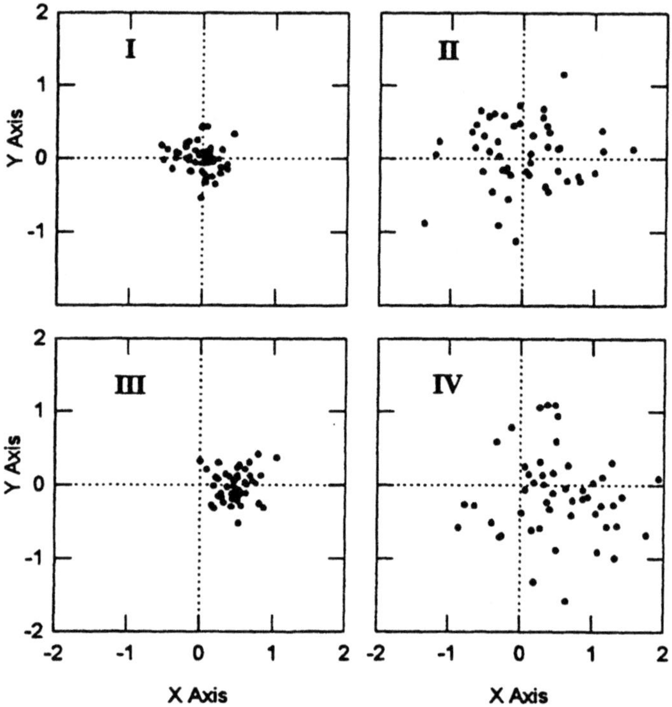

Fig. 2.1.a Application of the three key parameters for spatial accuracy and precision. The four examples shown are constructed using Monte Carlo methods. Each dot in these pictures is to illustrate the result of N=50 times applying a stereotactic localization procedure. All results are projected onto the two-dimensional x/y plane. In the four examples (*I - IV*) different uncertainties are demonstrated. The two examples on the *left* side (*I, III*) are constructed on a high precision basis, whereas in the two *lower* examples (*III, IV*) a systematic deviation of 0.5 mm in X was taken for the construction. *II* and/or *IV* may be typical of CT or MRI localization

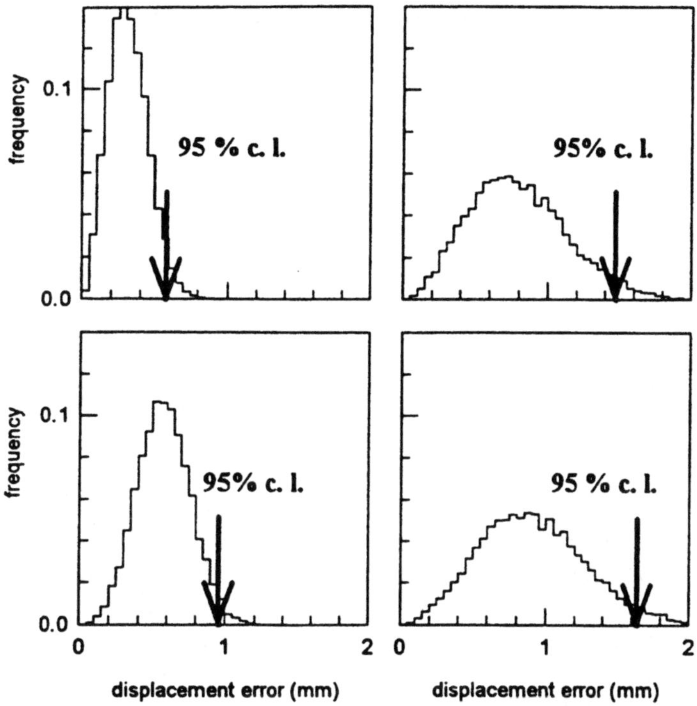

Fig. 2.1.b For each of the examples *I* - *IV* shown in Fig. 2.1.a, the corresponding frequency distribution of the displacement error r_i is given. The arrows indicate the value of $r_i + 2 \cdot s_r$, which is a suggested thickness of the security ring around the clinical target volume. The ring may typically have values between 1 and 2 mm for CT or MRI localization

2.2 Quality Assurance Aspects of Dosimetry and Treatment Planning

2.2.1 Background

2.2.1.1 Dosimetry

Dosimetry issues refer to beam data acquisition, absorbed dose calibration, and treatment planning verification. Beam data acquisition involves the measurement of percent depth doses (PDD) or tissue maximum/phantom ratios (TMR/TPR), profiles or off-axis ratios (OAR), and output factors. Small volume ionization chambers, silicon diodes, radiographic and radiochromic films, thermoluminescent detectors (TLD), diamond detectors, and scintillation detectors can all be used for relative dose measurements. In general, accuracy, precision, linearity and energy dependence of each of these detectors have to be carefully checked before meaningful measurements in radiosurgery dosimetry can be performed.

Dosimetric measurements at radiosurgical beams may require more attention in comparison to measurements in conventional radiotherapy: the dose may fall off significantly already within the sensitive volume of the detector; the detector may not correctly resolve the steep gradient; at very narrow beams, the lateral electronic equilibrium does not exist. The problems involved and methods of dose measurements at narrow beams are thoroughly described in the literature [3,8,10,12,14,16,26,34, 40,44,49,51,52,62].

For Gamma Knife based radiosurgery, beam data acquisition for one single beam is performed at a separate experimental setup for the following reason: at a Gamma Knife unit, basic measurements on one single beam could be performed only if the remaining 200 beams are extinguished. However, the transmission through each of the tungsten plugs used to extinguish beams in the Gamma Knife is always contributing to the dose delivered by the one single open beam. Although this transmission is only 0.2% for one plug, the transmission of all 200 plugs will amount to 40% and must therefore be subtracted. Because of the narrow tolerances to which the beam channels of the Gamma Knife are manufactured, an experimental setup which exactly simulates one beam channel of the Gamma Knife is used to provide the required basic data for one single beam. The experimental data have also been verified by Monte Carlo calculations. For each of the four helmets, these single beam data are prestored in the treatment planning system Gamma Plan.

Therefore, also for the purpose of calculated dose verifications, dose distributions should always be measured from results with all 201 beams unplugged.

2.2.1.2 Treatment Planning

The calculation of radiosurgical dose distributions is a three-dimensional problem that cannot be handled with the two-dimensional techniques used in standard radiotherapy treatment-planning computers. Therefore, it is essential that each medical center introducing clinical radiosurgery has access to a reliable three-dimensional target localization and treatment planning system. Radiosurgical treatment planning is based on stereotactic three-dimensional data obtained from CT, plain film angiography or MRI, in conjunction with stereotactic frames. Special care must be taken when using MRI because of spatial distortions which not only affect the target localization but may also adversely affect the dose distribution calculation. Target point localization is an important component of the treatment planning process. In addition, information about the patient's anatomical contours, which is obtained from a series of CT or MRI images, plays an important role in the calculation of the dose distribution. It provides the appropriate depth in tissue for each point-of-interest on the dose calculation matrix and enables an accurate calculation of the dose for each individual radiation beam. The patient's anatomical information, obtained with CT or MRI, is usually displayed on orthogonal planes (transverse, sagittal, or coronal) through the target. The calculated dose distribution should be directly superimposed onto the clinical CT or MRI images, so that doses are known not only within the targeted volume but also to sensitive or critical brain structures outside the target. In order to allow a quantitative comparison between calculated and measured dose distributions, the planning system should provide an option to display line dose profiles, at least along the principal axes through the target point.

A detailed description of treatment planning with the Gamma Knife is given in Chapter 3 of Philipps [39]. Briefly, the dose distribution from a Gamma Knife irradiation is calculated at each of 31^3 points in a cubic matrix, taking into account the dose contribution from all active sources. The matrix size and its location is individually chosen to cover the target volume and adjacent structures at risk. The dose distribution is tailored by means of helmet selection, by altering the number of target points and also their location and irradiation time or by extinguishing selectable beams. This planning process continues interactively until the dose distribution conforms with the planning target volume irrespectively of its shape. Subsequently, a treatment protocol is printed out, describing in detail the required settings of the Gamma Knife unit.

2.2.2 Key Parameters and Requirements

Dose accuracy specifications in stereotactic radiosurgery are the same as for conventional radiotherapy [1,11,27]. A further discussion from a clinical point of view is outlined in Appendix A. A relaxation of these dosimetry recommendations for very small fields of 10 mm in diameter or less is necessary because unresolved dose measurement discrepancies of several percent exist among different detectors and institutions.

The output of computerized treatment planning needs to be verified by measurements. These measurements fall into two categories:

1. Calculated relative dose distributions should be compared to three-dimensional relative measurements made with a high resolution detector such as film or possibly arrays of TLDs in a suitable phantom. The measured and calculated isodose surface in the high dose gradient region, i.e., 30% - 80% should agree spatially with each other at least within 1 mm. In the relatively flat high isodose region, i.e., 80% - 100% the relative dose for similar spatial points should agree within 3% - 4%.

2. Absolute dose measurements should be made at the reference setup point in a phantom for a simulated treatment and agree with

calculated absolute dose values to within 3%, if the detector response is normalized to that obtained in a reference field (e.g., 10x10 cm^2 at reference depth at a linac) on the same machine at the same time. If, on the other hand, a true absolute dose measurement is made using detector calibration factors in conjunction with a specific calibration protocol the agreement should be within 5%.

Note: A misadministration would occur if the calculated total administered dose differed from the prescribed dose by more than 10% of the total prescribed dose.

2.3 Quality Assurance Aspects in Treatment Delivery

2.3.1 Background

2.3.1.1 Gamma Knife Radiosurgery

There exist two Gamma Knife models (U and B). Differences in their spatial source distribution can be disregarded here. Movements in model U are accomplished by a hydraulic system whilst model B has electric motors. Both models comprise mainly three integrated parts: a stereotactic frame for target localization and head fixation, a dedicated treatment planning system Gamma Plan and the radiation shielded unit inside which the actual irradiation of the cerebral volume is performed.

The patient's head is surrounded by 201 small Co-60 sources during treatment (Fig. 2.2). Each source is individually collimated such that 201 stationary narrow gamma beams are precisely directed towards a common point, centrally located in the treatment unit. At this point, the so called unit center point (UCP), the radiosurgical target is aligned during treatment. No primary gamma beams exist in the treatment room.

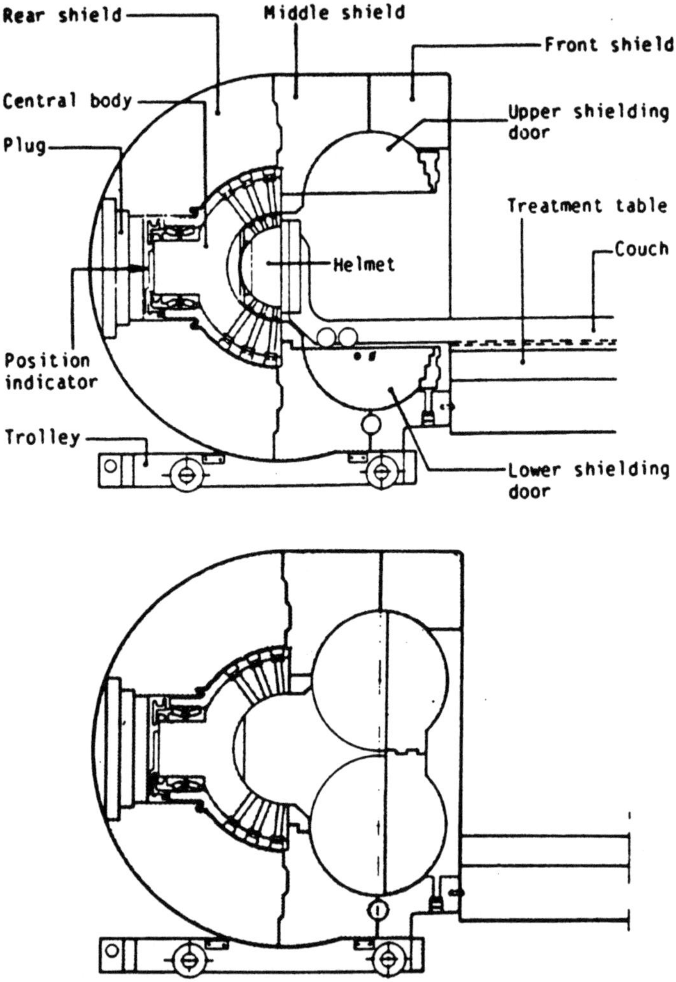

Fig. 2.2. Schematic drawing of the Gamma Knife. Above: Helmet in treatment position. Below: Unit switched off.

All 201 beam channels are manufactured to such narrow mechanical tolerances that they can be considered identical with respect to their radiation physical characteristics. As a consequence, one set of measured beam data is prestored in the planning software and used for all beam channels.

Final collimation of the beams is accomplished by 201 tungsten collimators located in an exchangeable helmet into which the patient's head is aligned and fixed during treatment. Four helmets are available, each containing a set of 201 identical tungsten collimators. The diameters of their circular apertures are 4, 8, 14 and 18 mm for the sets respectively, when projected on a perpendicularly oriented surface placed at the UCP.

2.3.1.2 Linac-Based Radiosurgery

The general availability of linear accelerators and advances in stereotactic localization techniques in the past decade have substantially contributed to the proliferation of medical centers offering stereotactic radiosurgery. Compared to other techniques, the linac-based method appears an attractive method, since only relatively minor modifications on a standard linear accelerator facility can make it useful for radiosurgery. However, the problem of achieving and maintaining the high accuracy needed for stereotactic radiosurgery in a linear accelerator facility is not as simple as with a Gamma Knife.

Linacs used for radiosurgery are isocentrically mounted and normally used for routine cancer therapy. A typical radiosurgery irradiation setup is shown in Fig. 2.3. The modifications and additions needed to adapt modern isocentric linacs to radiosurgery typically consist of: a set of additional collimators to define the small, circular radiation beams (typically 10 to 40 mm in diameter), a remotely controlled motorized couch rotation for one technique, brackets or a floorstand for mounting the stereotactic head frame, interlocked readouts for angular and height positions of the couch, special brakes to immobilize the longitudinal, lateral, and vertical couch motions during the treatment, and anticollision devices.

An important feature of linac-based stereotactic radiosurgery is the method of producing narrow high energy X - ray beams. Normally an additional collimator or a collimator housing is mounted at the auxiliary holder plate. A collimator housing allows various collimator inserts to produce narrow circular radiation beams. There have been some suggestions in the literature that the penumbra for narrow fields may be minimized by placing such collimators as far as possible from the source, and by constructing the collimator so that the defining edges of the collimator geometrically match the divergence of the radiation beam. Serago et al [49] investigated this problem for 6 MV X - rays and recommended that for beams with sizes 10 - 40 mm diameter there is no potential gain in decreased penumbra by using collimators with tapered holes compared to straight-drilled holes. They also have found minimal iinprovement in decreasing the penumbra when source to collimator distance was increased from 58.5 cm to 76 cm.

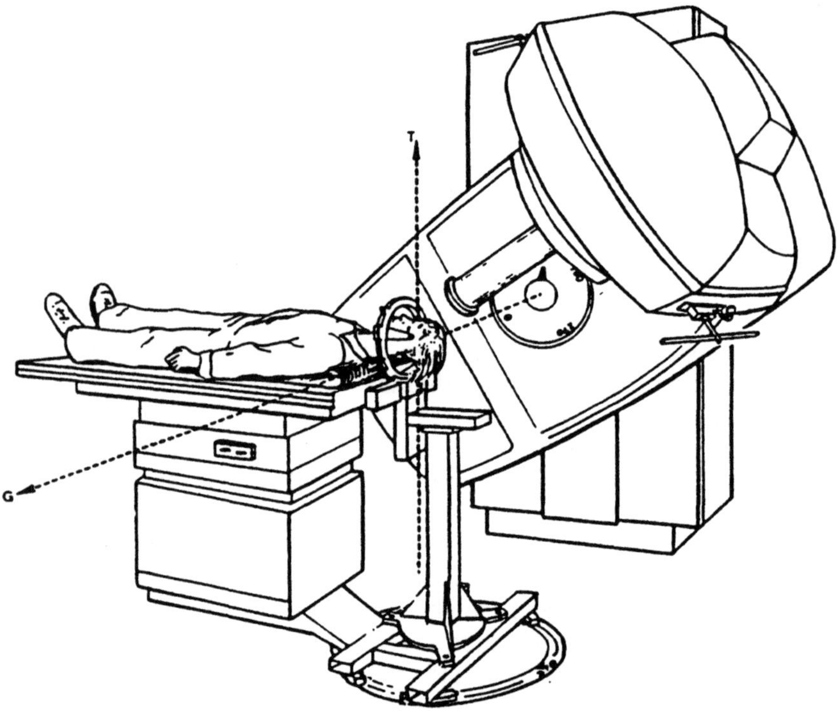

Fig. 2.3. A typical linac-based radiosurgery irradiation setup. This model has the Joint Center hardware configuration using a floor stand. The stereotactic frame is mounted to this floor stand. *Dotted lines* indicate the axes of rotation of the gantry (*G*) and the turntable (*T*)

Different linac-based radiosurgical techniques are characterized by the particular approach they take toward fulfilling the goal of delivering the prescribed dose to the target with a sharp dose fall-off outside the target volume. The techniques rely on specific combinations of the linac gantry and treatment couch/chair rotations, and currently fall into four categories: multiple non-coplanar converging beams [4,6,13,24,60], dynamic rotation [43], conical rotation [6,36], and miniaturized linac on a robot arm [2].

Two methods for patient setup have been developed for linac-based radiosurgery: (1) the stereotactic head ring or frame is attached to a floor stand which in turn is attached to the turntable of the treatment couch, (2), the stereotactic head ring or frame is fastened directly to the treatment couch. Each of these systems require some unique QA tests for patient setup. No matter which system is used, QA measures should be taken to ensure no inadvertent couch motion and to prevent collisions between gantry and couch, floor-stand or patient.

General quality assurance tests for normal radiotherapy [1] as well as radiosurgical quality assurance programs are published [15,33,56].

2.3.2 Key Parameters and Requirements

Accurate beam alignment is the most important requirement for the treatment delivery in stereotactic radiosurgery. Ideally, all beam axes intersect at one common point, the reference setup point (or UCP). In reality they do not. However, they should intersect within a small sphere. The diameter of this sphere can be used as a key parameter to characterize the quality of beam alignment.

For Gamma Knife based radiosurgery, the requirement according to the manufacturer's specification is, that all beam axes must intersect at the UCP within a sphere of 0.6 mm diameter. This specification is defined for the smallest beam size, i.e., the 4 mm collimator.

All current clinical linac-based radiosurgical techniques involve rotations of the gantry and the treatment couch. Therefore, accurate beam alignment depends particularly on the spatial location of couch and gantry axes. Their alignment and mechanical stability pose one of the most important problems in linac-based radiosurgery. Ideally, the central axis of the beam, and the axes of gantry and couch will intersect at one point independently of the rotational position of gantry and couch. This situation is shown in Fig. 2.4. In practice the axes do not intersect. The minimum requirement is that they should intersect within a sphere of diameter of less than 1.5 mm, but a diameter of 1 mm is generally desirable. Another useful key parameter, particularly for linac-based radiosurgery, is the distance of smallest approach between the gantry and the couch axes. This distance should definitely not be larger than 1.0 mm. Distances of 0.5 mm or less are achievable.

Accurate patient positioning is the second important issue in treatment delivery. For that purpose (1) the stereotactic coordinate system of the patient has to be aligned to the coordinate system of the radiation system unit, and (2) the patient's target point obtained from the treatment planning procedure must coincide with the reference setup point. This reference setup point as defined in section 1.2.2 is usually identified either by mechanical means or by a laser system. A possible deviation between these two points, the target point and the reference setup may occur. The same key parameters, $\bar{r}$, $\bar{r}_i$, and s_r, already introduced for the localization procedure, also apply for this step. Phantom test measurements (see tests 24 and 25) have demonstrated that positioning can be achieved with an accuracy of better than 0.5 mm [23].

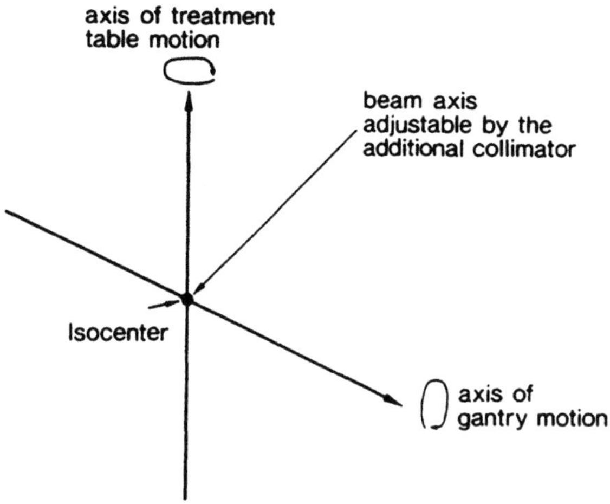

Fig. 2.4. Ideal alignment of couch and gantry to achieve a fixed mechanical isocenter: the central beam axis and the axes of gantry and couch rotation intersect at one point

3 Description of Recommended Quality Assurance Tests

The geometric miss of a radiosurgical target is considered the most likely serious error in a stereotactic radiosurgery treatment. For this reason, most of attention in this chapter on the quality assurance program will be directed towards minimizing this error.

The chapter is divided into the three sections: (1) stereotactic localization and immobilization, (2) dosimetry and treatment planning, and (3) treatment delivery, in the same way as the second chapter.

Each of these sections will begin with a survey of the major quality assurance that should be seriously taken into consideration. Following this, a series or recommended tests will be described. To facilitate the reading of the tests, each test is structured on the same plan: the **aim** which is addressed, the **method** used, the expected **results**, and the **type** of the test according to the level of application as given in Table 3.1. In this table the tests are classified as to their appropriateness for (A) initial tests of a system or a device, (B) for subsequent periodical tests, and (C) for every patient treatment. For Gamma Units, tests of type A normally have to be repeated if the radiation sources are moved or changed.

Table 3.1. Classification of test types

Type	Purpose of the test	Characteristics	Frequency
A: Initial status and suitability test	Determination of level of performance and decision on acceptance	Measurement of key parameter and base line values	Once at the beginning
B: Constancy test	Checking the reliability	Determination of consistency of key parameters	Periodically
C: Treatment test	Avoid mistreatment of a particular patient	Checking the performance before treatment	At every patient treatment

3.1 Stereotactic Localization and Immobilization

3.1.1 Major Quality Assurance Issues

- A rigid immobilization technique using skull fixation with pins does not guarantee the absence of all patient motion or organ motion.
- The extension arms of the head ring may be deflected with forces applied to the patient.
- The head ring may move during the radiosurgical procedures.
- Does the head ring and localization frame interfere with the image system?
- One should measure the accuracy and precision of the used localization procedure for known targets.
- Considerable geometric inaccuracies may occur when DSA or MRI are used for target point determination.
- Add a known target to the head ring or localization frame and test it with each patient localization procedure.
- The shape of the target volume cannot be determined from two (orthogonal) plain films.
- It is strongly recommended to determine the target point independently by two persons.
- The CT or MRI slice distance used has a direct impact on the precision of localization.

3.1.2 Tests

Tests 1 - 3 are concerned with single point localization using different imaging techniques for target selection. Tests to measure the accuracy and precision of localization alone, in the absence of any medical uncertainty, generally consist of using a device that provides a single test point (steel or plastic ball, tip of a needle). The device is rigidly attached to the stereotactic head frame. The position of the test point is obtained by mechanical means. These coordinates then serve as the "true" position.

Fig. 3.1. Example of providing a test target for stereotactic localization procedures. A target simulator with a steel ball representing a test target point is attached to the head frame. The "true" coordinates of the test point are mechanically determined

As an example, Fig. 3.1 shows a method as used with the Brown-Roberts-Wells (BRW) system. With this example, the "true" coordinates of the test point can be measured directly using the BRW phantom base. In general, angiographic localization was found to be very accurate. Therefore, if results of mechanical measurements are not available, "true" coordinates may be also obtained with a very good approximation by the angiographic localization method itself (given the method has been proven to be accurate previously).

The tests described in this section apply to the most commonly used imaging techniques for target identification:

Test 1: cerebral angiography (plain or cut film)

Test 2: computed tomography (CT)

Test 3: magnetic resonance imaging (MRI)

Magnetic resonance angiography, positron emission tomography, and single photon emission computed tomography are also under evaluation. Since their role in stereotactic radiosurgery is not yet established, specific tests for these methods and for digital subtraction angiography (which may have considerable inaccuracies) are not addressed.

With each of the following tests, a localization error, r_i, can be calculated at the i th measurement from:

$$r_i = \sqrt{(\Delta\, AP)^2 + (\Delta\, Lat)^2 + (\Delta\, Vert)^2}$$

where ΔAP, for example, is the AP difference between a mechanical (phantom base) measurement and that found through the treatment planning process. If this test is repeated a number of times, the mean and the variance of the localization errors will lead to the main key parameters, $\bar{r}_i$ and s_r, as introduced in Section 2.1.2 ($\bar{r}$ is obtained from a different formulation, see p. 7).

Note: $\bar{r}$ is a reasonable measure of systematic inaccuracies

$\bar{r}_i$ is a reasonable measure of the accuracy (the likely error at one single localization)

s_r is a reasonable measure of the precision

Test 1: Single Point Accuracy and Precision in Angiography

Aim. Inaccuracies in angiographic localization may be introduced by the stereotactic localization device or the image processing system (e.g. digitizing tablet, computational software, etc.). Before using a system on patients, these inaccuracies should be well understood.

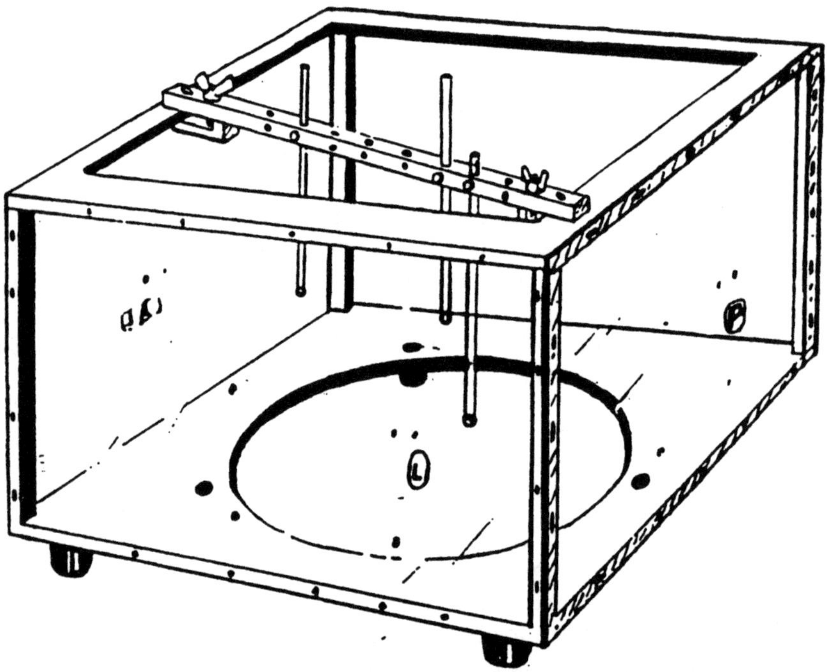

Fig. 3.2. Test of the angiographic (X - ray) localization procedure. Test targets are mechanically attached to the angiographic localizer, as used with the BRW system.

Method. Test targets (steel or plastic balls) are attached to the angiographic localizer frame. The coordinates of these test targets are then measured very accurately mechanically (the "true" coordinates). If the BRW system is being used, the phantom base is ideal for this purpose. Figure 3.2 shows this measurement being made on a test target attached to an angiographic box. For other systems a jig may have to be made. Next, the test target's coordinates are determined using specific localization software or the treatment planning program and compared to the coordinates measured mechanically.

Result. Plain film angiography is potentially very accurate and precise with typical reported values for $\bar{r}_i$ and s_r being 0.5 mm and 0.2 mm, respectively. It should be noted that two planar radiographs are inadequate to determine either the shape or size of the target volume.

Type. A

Test 2: Single Point Accuracy and Precision in CT Localization

Aim. Inaccuracies in CT localization may be introduced by the CT scanner, the choice of slice thickness, the stereotactic localization device, a tilt of the localization device to the scan axis or the image processing system (e.g. digitizing tablet, computational software, etc.). Before using a system on patients, these inaccuracies should be well understood.

Method. Test targets (steel or plastic balls) are attached to the CT localizer frame. The coordinates of these test targets are then measured very accurately mechanically (the „true" coordinates). If the BRW system is being used, the phantom base is ideal for this purpose. Figure 3.3 shows this measurement being made on a test target attached to a CT localizer frame. For other systems a jig may have to be made. Next, the test target's coordinates are determined using the treatment planning program and compared to the coordinates measured mechanically.

Result. CT localization is less accurate and precise than plain film angiography with typical reported values for $\bar{r}_i$ and s_r being 1.3 mm and 0.6 mm respectively.

Type. A, B

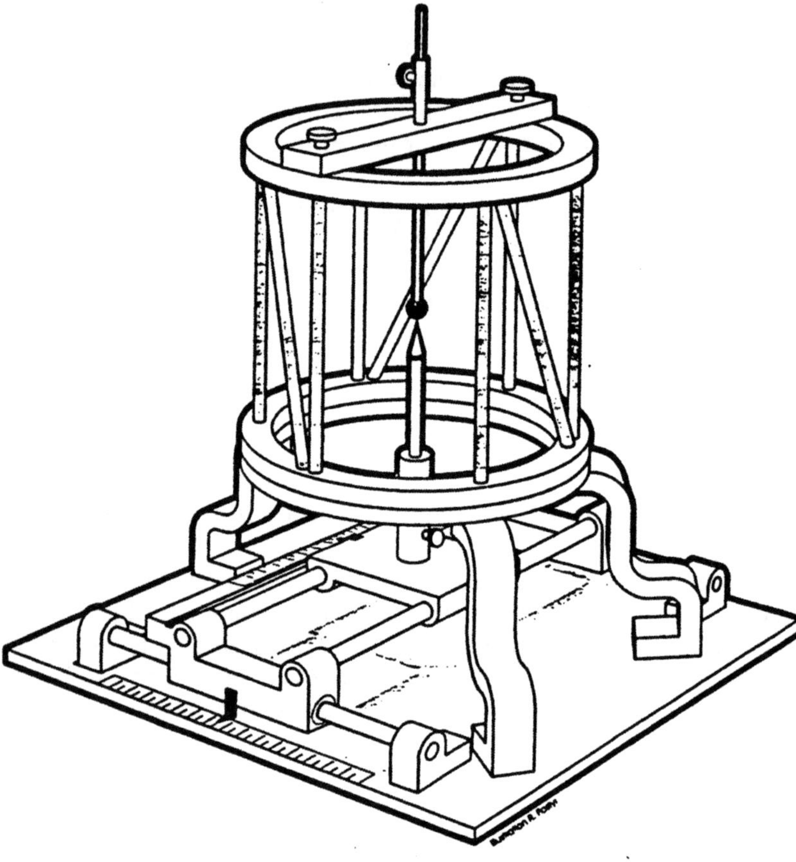

Fig. 3.3. Test of the CT localization procedure. In this example, the BRW system is used.. A steel ball representing the test target is attached to the CT localizer. The "true" coordinates of the test target point are measured directly using the BRW phantom base

Test 3: Single Point Accuracy and Precision in MRI Localization

Aim. Inaccuracies in MRI localization may be introduced by the MRI machine, stereotactic localization device or the image processing system (e.g., digitizing tablet, computational software, etc.). Due to possible image distortions, MRI localization is particularly susceptible to positional inaccuracy. Therefore, sizable inaccuracies can occur in localization with MRI. In addition, the MRI imaging system may not be stable. Before using MRI localization for patients, these inaccuracies should be well understood.

Method. Suitable test targets (e.g., objects filled with Gadolinium contrast media) are attached to the MRI localizer frame. Then the method is analogue to that with CT. A detailed method to correct for image distortions is by Schad et al. [46] and outlined in Appendix A3.

Result. MRI localization may be even less accurate and precise compared with CT. Values for $\bar{r}_i$ and s_r of 1.5 mm and 0.6 mm respectively were obtained. However, because this is not achievable at every MR unit, the reader is strongly urged to make his own comprehensive test.

Type. A, B

Test 4: "Known" Target Test During Localization with Patient

Aim. How does one know for sure that the three target coordinates (AP, lateral and superior-inferior) which will be used for a single isocenter treatment, for a particular patient, are correct? The problem is that one cannot. However, one possible localization test, which can at least check the localization process, involves placing a target of "known" coordinates external to the skull, but relatively near the internal target, during the angiographic, CT, or MRI localization procedure.

Method. The external "known" target is attached to or permanently embedded directly into the localization frame. The coordinates of the external target are determined mechanically. Following CT, angiography, or MRI, the coordinates of the external target are calculated using the same angiographic films, CT slices or MRI slices, computer hardware and software as are used to calculate the patient's treatment target coordinates. Only when these external target coordinates compare favorably with the "known" coordinates, does one proceed with the calculation of the patient's target coordinates.

Result. "Known" (mechanically measured) and calculated (treatment planning system) external coordinates should generally agree within 1.5 mm.

Type. C

Test 5: Head Ring Movement Test

Aim. The head ring may move between its initial placement and the treatment. Therefore, the position of the head ring with respect to the skull has to be measured and compared at placement and again just before treatment. A similar test is needed when repositional head rings or masks are used for fractionated stereotactic radiotherapy.

Method. After the head ring is secured, the stereotactic biopsy arc is attached to the head ring (Fig. 3.4). (An alternative jig may be constructed which is more convenient for this purpose.) Using the arc system and a "biopsy" probe, three or four points on the scalp are marked at specific arc coordinates (usually angles and depths). Choose "easy" numbers for the angles, since the only constraint is that the points be distributed. Just prior to treatment, the biopsy arc is again attached to the head ring with the patient's head in the same position as before. The three or four marked scalp points are then confirmed to be at the same coordinate values, thus establishing that the head ring has not moved.

Result. At the time of treatment, a "biopsy" probe set to the original coordinates should be within 1 mm of the original scalp mark.

Type. C

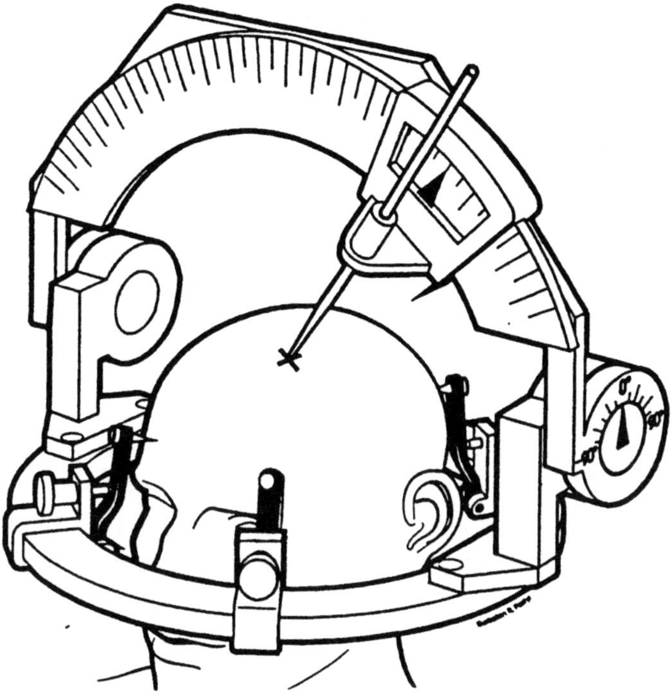

Fig. 3.4. Head ring movement test

3.2 Dosimetry and Treatment Planning

3.2.1 Major Quality Assurance Issues

- Issues involving detector size need to be examined.
- Set the jaws always at the same maximum acceptable jaw position.
- All depth dose data (PDD, TMR, TPR) should be normalized to one fixed point in depth.
- For very small fields (below 10 mm diameter) special dosimetry problems may exist.
- For each relative and absolute dose measurement with film, a calibration performed and processed at the same time is necessary.
- Compare your measured data set (TPR, OAR, OF) with at least one other set measured at another institution for a beam of the same energy.
- Verify that the treatment program produces accurate isodose distributions.
- Verify that the treatment program calculates correctly the monitor units, i.e., the absolute dose delivery.
- For each patient, an independent check (usually a hand calculation) of the monitor units is strongly recommended.

3.2.2 Tests

Test 6: Detector Size Appropriateness

Aim. The most important criteria on the choice of a detector is its size in relation to the beam size. To minimize size effects, detectors of the appropriate size should be used.

Method. For central axis measurements (e.g., PDD, TMR, TPR, and output factors) the detector (generally an ion chamber) should have small dimensions compared to the diameter of the field, i.e., the diameter of a detector (as viewed from the radiation source) should be less than 1/3 of the field diameter. For most of the radiosurgical beams, cylindrical ionization chambers with 0.1 cc volume are adequate. For small field sizes, it is advisable to turn the axis of the detector parallel to the beam central axis. For field sizes smaller than 12 mm to 15 mm diameter it is recommended to average measurements from several detectors (e.g., ionization chamber, diode, diamond detector, TLD). This is particularly important for output factor measurements.

Dose distribution measurements involving sharp gradients (for both single or multiple beams) require detectors with a high spatial resolution. Film, properly calibrated, works well for this purpose provided it is scanned with small apertures (1 mm). Small ion chambers, TLDs and diodes may also be used. If a sufficiently small detector is not available, the influence of the detector size on measured dose profiles can be corrected [10,14,44,51].

One detector is specifically meant: there is a natural diamond detector whose suitability for dose measurements in narrow beams was demonstrated in recent publications [17,49,58, 59]. It was shown that the good spatial resolution and high sensitivity allow one to position the detector precisely on the beam axis and to determine the absorbed dose to water.

In terms of the cavity theory the absorbed dose to water D(w,u) for the photon beam quality *u* would be as follows:

$$D(w,u) = M_u \cdot N_{w,cal} \cdot \frac{P_u}{P_{cal}} \cdot \frac{S(w,c)_u}{S(w,c)_{cal}}$$

where

u refers to the measurements at radiation quality u

cal refers to calibration in a water phantom using Co-60 beam

w means water

c means carbon

M is a measured charge from the diamond detector

$N_{w,cal}$ is absorbed dose to water Co-60 calibration factor

P is a perturbation correction factor

$S(w,c)$ is the stopping power ratio of water to carbon.

As a diamond detector can be considered as a disc-shaped detector, the ratio of perturbation factor can safely set equal to unity [55]. The variation of the stopping power ratio water to carbon for photon energies from Co-60 to 4 -10 MV is very small compared with the water to air stopping power ratio and can be set also to unity with good approximation. Then the absorbed dose in a radiosurgical X-ray beam can be calculated as a product of measured charge and the Co-60 absorbed dose to water calibration factor $N_{w,cal}$. The diamond detector provides measurements for field sizes down to 3 - 5 mm, as the thickness of the diamond detector plate is 0.2 - 0.3 mm and the diameter of the sensitive area is 3 mm [58].

Result. The same dosimetrical parameter should be measured using several different techniques and detectors. Then the results should be compared to confirm accurate measurements and to identify errors and inadequate measurement techniques.

Type. A

Test 7: Accuracy of Detector Positioning

Aim. During the measurements, careful attention has to be paid to the spatial positioning of the detector. This is especially true for central axis measurements like depth dose measurements where the detector moves along the central axis.

Method. To achieve coincidence of the center of the sensitive detector volume with the beam axis, the positioning of the detector should be done by measuring X- and Y-profiles at two depths. The maximum reading at a given depth is the one used to center the detector.

Result. For each depth the maximum value of the measured X- and Y-profiles should be the same. Moreover, the central axis deviation of all profiles should be also the same. A wrong positioning may be corrected by adjusting the phantom and/or the gantry angle (for linacs).

Type. A, B

Test 8: Single Beam Model for Treatment Planning

Aim. The beam model used in the treatment planning program must adequately describe the absolute dose distribution of a single stereotactic beam. Dosimetrical data have to be measured for that purpose.

Method. Because the role of scatter is diminished for small beams, beam profiles are essentially independent of depth (after magnification adjustments). For this reason, the dose at any point in a fixed beam can be obtained quite simply from a limited amount of data. Basically, one needs the tissue-phantom ratios, output factors, and beam profiles:

$$D(c,d,r,R) = TPR(c,d) \cdot OAR(c,r') \cdot OF(c) \cdot \left(\frac{SID}{R}\right)^2 \cdot CAL \cdot MU$$

where

c = collimator diameter

d = depth

r = radial distance from central axis to point of interest

R = distance from source to point of interest

$TPR(c,d)$ = tissue phantom ratio for the collimator diameter c and at depth d. Tissue-phantom-ratios can be obtained from measured central depth dose data using:

$$TPR(c,d) = \frac{PDD(c,d)}{100}\left(\frac{SSD+d}{SSD+d_{max}}\right)^2$$

$OAR(c,r')$ = off axis ratio
Beam profiles are essentially independent of depth when measured at the same distance from the source. This empirical observation suggests that the radial distribution of a given beam can be adequately described by a single profile, preferable at isocenter distance from the source, SID. The value of r' in OAR is calculated from:

$$r' = r \cdot SID / R$$

Beam profiles can be most simply measured by film.

$OF(c)$ = relative output factor for a collimator c

MU = number of monitor units

CAL = calibration factor in Gy/MU at the reference point. It should be measured with ionization chambers.

Following this, the measured dose and the dose obtained from the beam model need to be compared at representative points within a given beam.

Result. (See Sect. 2.2.2)

Type. A

Test 9: Treatment Planning Verification

Aim. The output of computerized treatment planning needs to be verified by measurements.

Method. Since a thorough test of the treatment planning system is much more elaborate compared to the tests presented so far, only few aspects can be addressed here (for more details, see Appendix B).

The two most important dosimetrical items that need to be verified are (1) the correct calculation of monitor units (or time) to deliver a certain dose, and (2) the correct calculation of the radiosurgical dose distribution. To measure dose distributions several methods can be applied. Profile measurements [41] can be performed in a spherical or humanoid solid head phantom using films or TL dosimeter with corrections for their finite length [34] where an accuracy up to ± 2% can be achieved. The use of film cutouts irradiated in a black polystyrene phantom for treatment planning verification was studied by Bardash et al. [5]. They used a 1 mm aperture for film scanning and demonstrated the suitability of film technique for verification procedure.

Result. (See Sect. 2.2.2)

Type. A

3.3 Treatment Delivery

3.3.1 Major Quality Assurance Issues

General issues:

- The most important test of treatment delivery is a system integration test for geometrical accuracy using a target of unknown position in a head phantom.
- Test the head ring for movement during the entire radiosurgery procedure.
- Verify that target point coordinates will be actually treated.
- The torque applied by the patient's weight may cause deflection of the head ring from its intended position prior to each therapeutic irradiation.
- Check all details of treatment delivery by following an extensive check list covering all these details.

Issues specific to Gamma Knives:

Check whether

- Correct collimator helmet is selected.
- Correct collimators in the helmet are exchanged with plugs to achieve protection of the lens and/or tailoring of the dose distribution.
- Correct coordinates are set on the stereotactic frame and trunnions.
- Correct treatment time is set.

Issues specific to linear accelerator facilities:

- Mechanical alignment of gantry and treatment couch rotation axes.
- Mechanical alignment of the central beam axis of the radiosurgical beam.
- Minimizing the best compromise center sphere where all three axes come closest together.
- Determining the reference center point of that sphere.
- Conformation of the location of the reference setup point.
- Correct alignment of all three laser beams is critical.
- Hands-on measurement of target positioning just prior to treatment using lasers and rulers.
- Absolute immobilization of couch before treatment.
- Protection against gantry collisions.

3.3.2 Tests

Especially at the step of treatment delivery one should never forget that a potentially detrimental radiation dose is delivered to a human being in an area that is vital. Therefore, all steps of the radiosurgical procedure should always be checked independently by a second person.

The tests described below in this part are either appropriate for Gamma Knife radiosurgery and/or linac-based approaches to radiosurgery (using couch or floorstand). Therefore, the following section is divided into three parts applying specifically to (1) Gamma Knife based radiosurgery, (2) linac-based radiosurgery, or (3) tests independent of a method.

3.3.2.1 Tests for Gamma Knife (Model B) Based Radiosurgery

Test 10: Air Kerma Rates in the Treatment Room

Aim. Air kerma values are provided by the manufacturer for points located 60 cm from the surface of the unit (1 m from the closest source) and also for points at 1 m from the floor level and distributed in the treatment room. These later values are given for different operational conditions of the Gamma Knife. At least at some statistically chosen points, the given values should be verified.

Method. Measure the air kerma rate with a calibrated survey meter designed to measure photon radiation in the energy range of 1 MeV and air kerma rates in the range of mSv h^{-1}.

Result. The measured data should not exceed the data provided by the manufacturer of the Gamma Knife by more than 20%.

Type. A

Test 11: Wipe Test

Aim. Although it is extremely unlikely that radioactive material will leak through the encapsulation of the radiation sources this should be ascertained by wipe tests.

Method. Wipe the outer surface of each collimator helmet with a small piece of soft tissue wet with alcohol. Measure for several hours the tissue in a shielded well counter connected to a spectrum analyzer.

Result. The measured contamination must not exceed the maximum values recommended by the governmental authority for each country.

Type. A, B

Test 12: Geometrical Accuracy Related to the Treatment Unit

Aim. It is imperative to know how well the radiation dose can be delivered geometrically to a pre-determined point in the stereotactic space. For that purpose the geometrical accuracy (discrepancy) between the location of the point of dose rate maximum in the "non attenuated" radiation field of the unit and the location of the UCP within the treatment unit must be determined. The cause of a possible discrepancy may be an inaccuracy (1) of the collimator helmet, or, (2) of the trunnions used to position and immobilize the stereotactic frame. These trunnions, although made of steel and very rigid, may be damaged by careless handling and hence introduce geometrical errors in the positioning of the target point to the UCP. Therefore two different methods are described referring to each of these causes.

Method 1. Location of the Radiation Focus. This method is based on film dosimetry. An accurately machined cylinder shaped film holder, made of aluminum, is delivered with the Gamma Knife. This film holder contains, in addition to a small space for the film, also a sharp spring loaded needle. The film holder is aligned and fixed in the 4 mm collimator helmet by means of the trunnions, so that the position of the needle tip exactly coincides with the UCP (both trunnions locked at 100.0). After exposure the location of a narrow hole, pierced through the exposed film, is compared with the location of the center of the density distribution. By combining results from measurements in three perpendicular directions, the distance between the UCP (mechanically defined) and the target point (radiological defined) are determined (Fig. 3.5). This distance defines, within the methodological errors, the accuracy with which a predetermined point can be aligned in the unit.

Result. The tolerance for this measurement is specified to 0.5 mm. In all units tested so far this distance has been found to be between 0.1 mm and 0.3 mm. The results from this experiment should be archived as reference for subsequent measurements.

Type. A

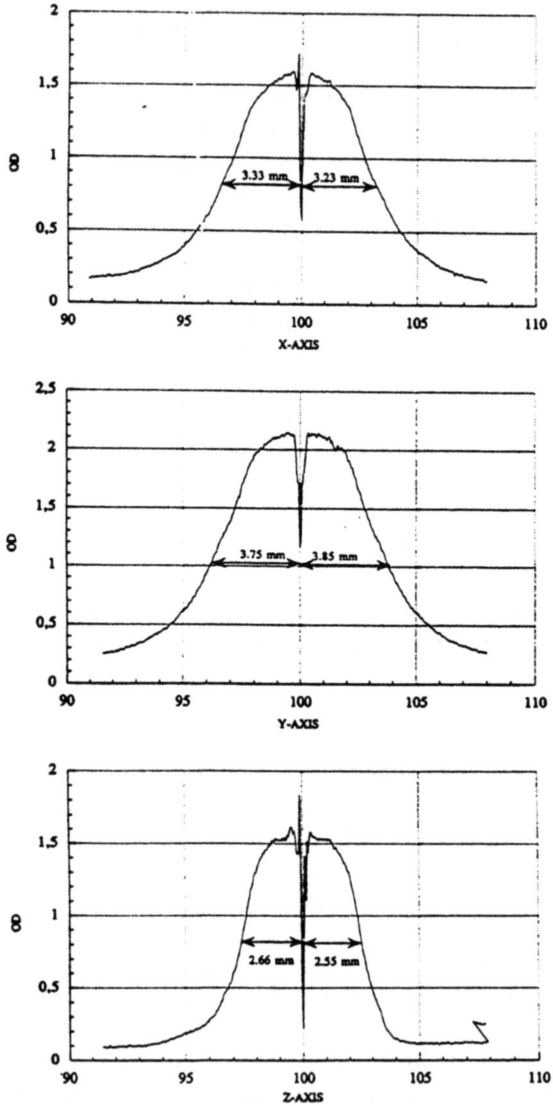

Fig. 3.5a-c. The location of the geometrical defined UCP is seen as a narrow valley in the optical density (OD) distribution from the 4 mm collimator helmet. The asymmetry of the OD distribution is measured at FWHM of the OD distribution in all three perpendicular planes. From these deviations the accuracy of the location of the radiation focus is calculated

Method 2. Mechanical Test of the Trunnions. A very accurately machined tool (cross bar) is delivered with the Gamma Knife. When the tool is attached to the helmet by its four screws, the ball of each trunnion fits into a position in the tool. Before reading, lock the trunnions by their locking screws.

Result. The coordinate X = 100.0 ± 0.1 should be read on the scale of each trunnion.

Type. A, B

Test 13: Dose Delivery and Geometrical Dose Distribution Related to the Treatment Unit

Aim. Dose delivery and the exact geometrical shape of the dose distribution at and around the focus depend on a number of parameters such as the strength of each radiation source, its exact alignment in relation to the collimator system, the tolerances to which the collimators are manufactured as well as their exact alignment. These parameters may affect the dose distribution not only in space but also in time. Therefore the actual dose distribution at and around the focus must be assessed and compared with that used at the treatment planning program.

Method. A special polystyrene phantom 160 mm in diameter is delivered with the Gamma Knife (Fig. 3.6). This phantom can be positioned and immobilized between the trunnions. Cassettes of different design can be inserted in the phantom allowing measurements with film, TLD or small ionization chambers. Because of the very high dose gradients existing in the radiation field to be measured, use of film dosimetry (radiographic or chromatographic) is recommended.
For each of the collimator helmets two films are consecutively irradiated in the spherical phantom under mutually perpendicular orientations. Density profiles are measured from these films along the stereotactic main coordinate axes of the frame. These profiles are converted to relative dose profiles and compared with those calculated. As the hypothetical dose profiles are calculated assuming ex-

act beam alignment and identical source strength it is practical to use these profiles as reference data.

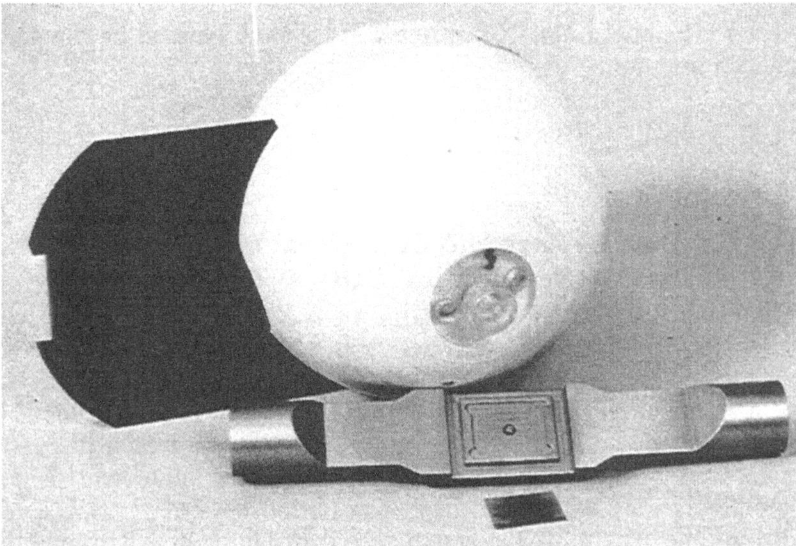

Fig. 3.6. The special polystyrene phantom used to measure dose and dose distributions

Result. The experimental and the hypothetical dose distribution normalized to dose maximum are compared by measuring the width at the 50% level (Fig. 3.7). The measured width must agree within ± 1 mm with that of the corresponding calculated profile. The data should be stored as reference for future subsequent measurements.

Type. A, B

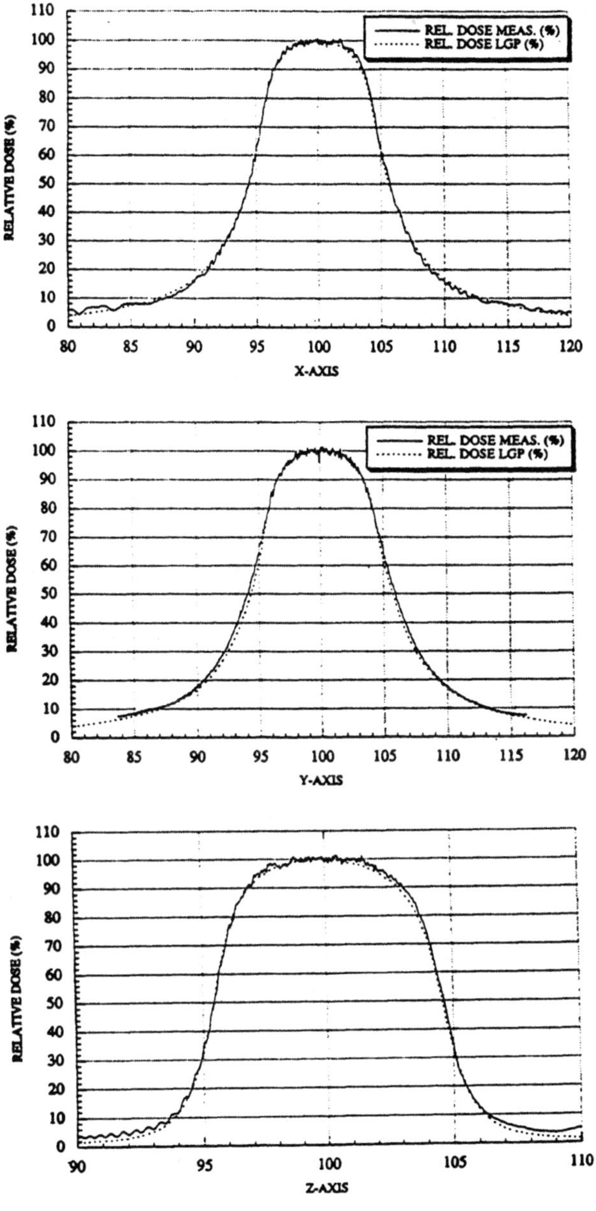

Fig. 3.7a-c. Comparison of profiles of a measured relative dose for the 8 mm collimator helmet with profiles calculated by the treatment planning system (LGP)

Test 14: Absolute Dose Calibration

Aim. In a consistent manner, all dosimetrical characteristics of the Gamma Knife can be described in terms of relative dose (for that purpose film dosimetry is normally used). The only additional dosimetrical data, which is required for the purpose of treatment planning and which is individual for each unit, is the actual absolute dose rate. The dose rate and the date of its measurement is needed to calculate the treatment time, corrected for source decay for each "shot" (irradiation).

Method. It is recommended to measure the absolute absorbed dose rate with an ionization chamber method traceable to the national standard laboratory. Of course, the dose must be measured in a radiation field that is homogeneous over the entire sensitive volume of the detector. Since the smallest commercially available ionization chambers are in the order of approximately 0.1 cc, such measurements in practical terms should be performed at the UCP when the largest 18 mm collimator helmet is used.

For this collimator helmet, the dose rate must be measured at the center of the spherical phantom (Fig. 3.6) with all sources active. The dose rate for the remaining three smaller collimator sizes (4 mm, 8 mm, and 14 mm) are derived from this measurement by corresponding output factors relative to the 18 mm collimator helmet which are also called relative helmet factors. These factors can be measured with any relative dose measurement method. For comparison, such data (measured and calculated) are also provided by the manufacturer as prestored data in the treatment planning program.

Result. The dose rate for the 18 mm collimator helmet shall be > 3 Gy/min at the time of the initial acceptance test.

Type. A, B

Test 15: Treatment Position of Collimator Helmet

Aim. In each of the four collimator helmets there are two adjustable micro switches sensing the exact position of the helmet with a tolerance of $\pm$ 0.1 mm when in treatment position. If the helmet for some reason does not reach this position, the treatment is automatically aborted. The treatment couch will be ejected within 2 seconds. In order to avoid unnecessary interruption in the Gamma Knife procedure, the correct adjustment of these switches should be checked on a regular basis.

Method. A metallic ring and a set of indicator lamps is delivered with the Gamma Knife. The ring simulates the contact surface for the helmet when in treatment position. The indicator lamps, when connected to the connectors in the helmet, show the status of the micro switches of the helmet. Place the ring over the helmet and connect the lamp set via its cable to the helmet.

Result. Both green lamps should only be lit. Shim the ring with a 0.15 mm thick piece of metal. The red lamps should only be lit. Adjust the micro switches if necessary.

Type. A, B

Test 16: Safety Features in the Gamma Knife

Aim. The proper function of all safety functions of the Gamma Knife must be verified. These functions are generally not activated during normal treatment conditions and should therefore be tested on a regular bases.

Method. Prior to the first treatment of each day, a dummy treatment should be made. Its purpose is to ensure the proper functioning of the safety features of the Gamma Knife. The most practical way to perform these tests is to provoke the initiation of safety function during a dummy run, i.e., to simulate a treatment during which safety functions are artificially triggered.

Initiate a one minute treatment. When the couch has reached its treatment position check the following points:

1. Failure of hospital power distribution:

Switch off the main power to the Gamma Knife by the hospital's power breaker. All functions of the unit shall remain active. Continue directly with test 2.

2. Treatment interrupt function:

   Interrupt the treatment. The unit shall go automatically to its idle status. Continue with test 3.

3. Freeze all movements of the Gamma Knife:

   Turn on the hospital's power again and restart treatment. Freeze all movements of the unit at an arbitrarily chosen time during treatment. The unit must instantly freeze all movements. Continue directly with test 4.

4. Function of automatic couch withdrawal and emergency:

   When the timers have reached their pre-set value the timer REMAINING time must continue through zero time and continue with negative values. The timer ELAPSED treatment time must continue also when the pre-set treatment time is exceeded.

Type. A, B

3.3.2.2 Tests for Linac-Based Radiosurgery

Test 17: Wobble of Gantry or Couch Rotation Movement

Aim. During a full rotation a wobble of the gantry or couch may occur. Gantry sag is another possible inadvertent movement.

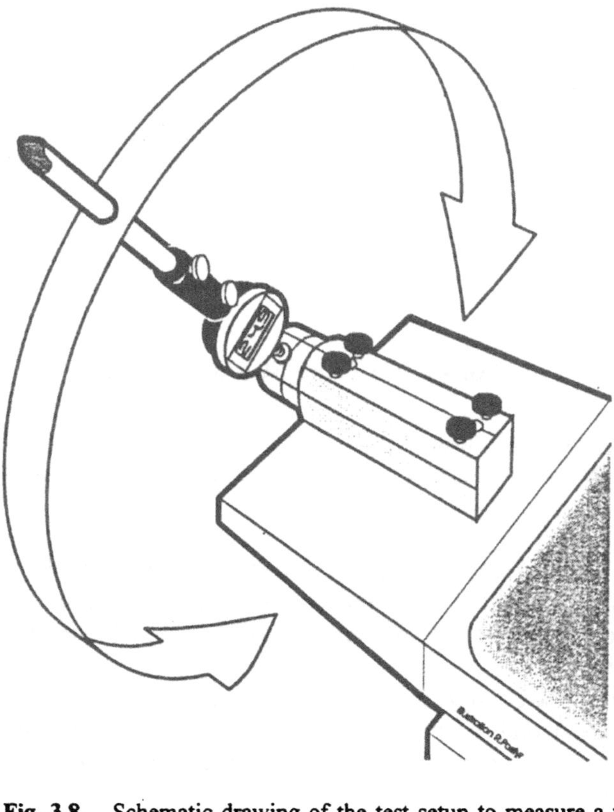

Fig. 3.8. Schematic drawing of the test setup to measure a wobble of gantry or couch movements

Method. A cylinder of 2-3 cm diameter, a mechanical measuring device such as an indicator gauge, and a rigid beam pointer are needed. For gantry wobble measurements, the cylinder is horizontally attached to the table with its axis aligned along the nominal gantry rotation axis; the indicator gauge is fixed at the beam pointer attached to the secondary, radiosurgery collimator holder (Fig. 3.8). The indicator gauge then records each deviation perpendicular to the axis during one rotation. Repeated measurements will help to verify that the cylinder is perfectly adjusted along the gantry rotation axis. The measured remaining deviations indicate a wobble during gantry movement.

For couch wobble measurements, the cylinder is vertically and rigidly attached at the gantry (via the beam pointer) and the indicator gauge to the couch. The cylinder axis has to be adjusted to the couch rotation axis. The same procedure is then performed.

Result. The maximum deviation between the perpendicular and the rotational axis must be less than 0.5 mm. Otherwise the facility should not be used for radiosurgery.

Type. A

Test 18: Alignment of Gantry and Couch Rotation Axes

Aim. If a linac is used for radiosurgery it is imperative that the gantry and couch axes of rotation very nearly intersect (within at least 0.5 mm).

Method. A test is described which can measure the length of the shortest line connecting the gantry and couch axes of rotation. The length of this line is a measure of the degree to which these axes "miss" intersecting.
Obtain a horizontal platform with one and preferably two orthogonal micrometer drives. Attach a sharp vertical pointer to the moveable platform and place this moveable pointer system on the couch with the point at the nominal isocenter (Fig. 3.9).

Next, attach a second rigid pointer to the secondary, radiosurgery collimator holder. If possible, this pointer should be concentric with the central beam axis and have its tip placed at the nominal isocenter. Start with the gantry in the upright position (beam aimed down). Adjust the couch pointer using the micrometer drives such that its tip remains fixed under couch rotation. The collimator pointer is a handy reference.

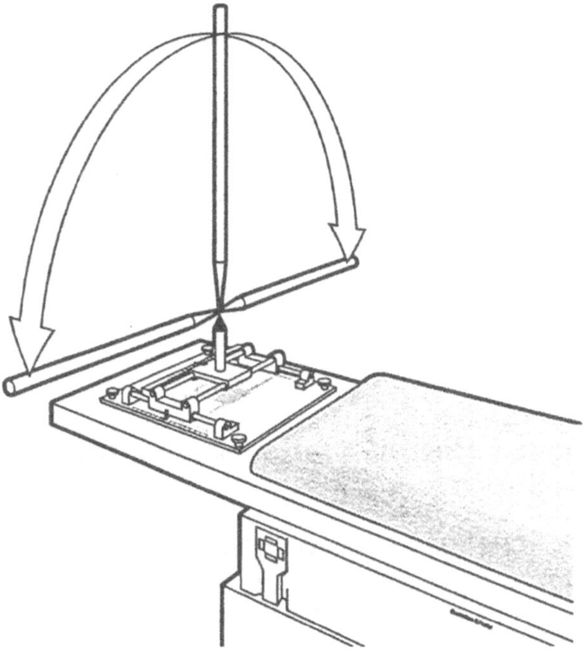

Fig. 3.9. Measuring the distance between gantry and couch rotation axes

After aligning the couch pointer, be careful not to bump the couch during the remainder of the procedure. Rotate the gantry to its 90° position. Adjust the gantry pointer so it is even in the left-to-right direction with the couch pointer. The gantry pointer may be positioned slightly toward the linac stand to avoid contact. Now rotate the gantry to the 270° position again being careful that the pointers do not collide. Most likely the left-to-right direction is slightly different. Adjust the gantry pointer to the "left-to-right" midpoint between these two pointer positions (gantry 90° and 270°). Repeat the process until the gantry pointer does not move in the left-to-right direction under the 180° gantry rotation between left and right lateral positions. The tip of the gantry pointer at either 90° or 270° indicates the left-to-right position of the axis of gantry rotation (the height of this axis does not matter for this measurement). With the gantry at 90° or 270°, move the couch pointer left or right with the micrometer drive until the two points are coincident in the left-to-right direction noting the measurement in mm of travel. The amount of this movement is an exact measure of the closeness of the intersection of gantry and couch axes of rotation.

Result. It is generally felt that gantry and couch axes should intersect within 0.5 mm. The great majority of linacs used for radiosurgery fulfill this condition. Larger deviations than this suggest the need to realign these axes by either moving the couch or the machine.

Type. A

Test 19: Measuring the Effect of Gantry "Sag" on the Beam Axis

Aim. The central axis of the beam is known to "walk" along the "IN-OUT" direction (parallel to the gantry axis) as a function of gantry position. The range of this motion should not exceed given tolerances.

Method. A test is described to measure the movement of the beam central axis along the gantry axis direction.

The same apparatus and a similar setup should be used as that for measuring the couch-gantry axes of intersection previously discussed. Rotate the gantry 180° from its "upright" position (beam aimed downward) to its "lowest" position (beam aimed upward). Use the couch position pointer and its micrometer drive mounted sideways and beyond the end of the couch as a reference to measure the range of movement of the gantry pointer in the IN-OUT direction (along the gantry axis of rotation) as a result of gantry rotation. This is the so-called gantry sag effect.

Result. On most linacs, the range of motion of the central beam axis along the axis of gantry rotation, as a result of gantry rotation, is 0.6 to 0.8 mm. This level of motion is considered acceptable. Major mechanical work would have to be done to a linac gantry to reduce this effect.

Type. A

Note: Following tests 18 and 19 it is recommended that the reference setup point at a linear accelerator facility (i.e., the location at which the target point,is placed through its coordinates) should be located at the point on the axis of couch rotation at the position nearest the axis of gantry rotation. The central axis of the beam should intersect the gantry axis of rotation at a point nearest the axis of couch rotation. Because of gantry "sag" this can only be achieved at one gantry position. The logical choice is the intermediate or lateral gantry position.

Test 20: Secondary Collimator Alignment

Aim. The collimator alignment represents a further critical step to achieve a focal dose distribution with a linear accelerator. The collimator usually is placed onto the linac tray holder, which is attached to the linac head. It must be adjusted in such a way that the central collimator axis defining the beam axis exactly coincides with the line connecting the radiation source (beam spot) to the reference setup point. If the gantry and couch axes do not intersect at one point, see Note above. The prerequisite for this adjustment is that gantry and couch axes have been well aligned previously.

Methods. Two methods are described to test for proper secondary collimator alignment: (1) a general method using a mechanical pointer and (2) a method using a dot producing, vertical ceiling or the field size lamp with a circular collimator inserted.

Method 1. Two pieces of alignment equipment are necessary. Firstly, a mechanical pointer and its holder should be machined such that the pointer axis is coincident with the central axis of the collimator insert. A perfectly straight drill rod and a plastic holder turned on a lathe to the size of a collimator insert with a central hole to accept the rod works well. The rod should be adjustable in length and reach the isocenter. Secondly, a machined disk, with a lead plug (about 3 mm in diameter and 1 cm long) in the center, should fit at the proximal end of the secondary collimator holder at a maximum distance from the collimator insert. There must be separation between the two for this test to be sensitive to misalignment. The collimator holder and the disk share a common symmetry axis having both been turned on a lathe. A properly aligned collimator should simultaneously satisfy the following two conditions:

a. The tip of the mechanical pointer should not move in the plane of gantry rotation as the gantry rotates through 360°. The tip should be "even", i.e., same IN-OUT direction with the couch axis of rotation when the gantry is in the lateral position.

b. The pointer is removed and a 1 cm diameter collimator insert is installed. Additionally, the proximal disk with the lead center marker is installed. A radiograph is taken with film located immediately distal to the collimator insert. The lead marker should appear exactly in the center of the 1 cm diameter beam.

Result. In general with patience a collimator can be adjusted, so that each test demonstrates misalignment of less than 0.3 mm.

Type. A, B

Method 2. Alignment can be nicely performed with a dot producing, vertical ceiling laser. The couch table top is raised to the isocenter and the gantry positioned off-vertical, so that the dot produced by the vertical ceiling laser to indicate the couch rotation axis is clearly visible on the table top. The position of the dot is marked on the table top, the gantry placed vertically, and the collimator position is centered with the help of the light-field localizer lamp around the mark on the table top representing the couch rotation axis. The collimator is then immobilized.

Alignment is tested with three exposures of a radiographic film on the treatment couch at the isocenter. The first test is done with the film in a horizontal position irradiated with two vertical-opposed beams (gantry at 180° and 0°), the second test, with the film in a vertical position irradiated with two horizontal parallel-opposed beams (gantry at 90° and 270°), and the third test, with the film in the horizontal position irradiated with a vertical beam (gantry at 180°) at several couch angles between 75° and -75°. The first two test procedures confirm the alignment of the collimator with the isocenter, whereas the third test confirms the alignment of the couch rotation axis. Such films may also serve to document the actual alignment for the patient's record file.

Result. The alignment of the images should fall within 0.3 mm.

Type. A, B, C

Test 21: Beam Spot

Aim. A stable beam focus or spot is essential to preserve the geometric integrity of the stereotactic treatment system. The beam spot may change with gantry angle or with time. The aim of this test is to investigate the beam spot. This test, however, is not considered as a routine test.

Method. A beam spot camera as described by Lutz et al. [32] and applied as given by Tsai et al. [56] can provide two pieces of information: (a) it can image the shape of the beam spot. (b) If the camera is fixed with respect to the jaws of the gantry, it can detect movements of the beam spot relative to the central axis of the camera.

A beam spot camera with a film at its end is rigidly mounted to the gantry face. At gantry angles of 90°, 180°, 270°, and 360°, a film is exposed. These exposures are repeated with the beam spot camera rotated 90°. In this manner the movement of the beam spot and its full width at half maximum is measured in two orthogonal directions. The films are then scanned in a densitometer which provides high resolution.

Result. Acceptable values for beam spot movement are in the order of 0.2 mm.

Type. A

Test 22: Marking the Reference Setup Point with a Laser System (Heidelberg Test)

Aim. The reference setup point, though invisible, is a well defined fixed point in the facility's coordinate system. Accurate knowledge of its location in space is extremely critical with respect to patient positioning and thus the total accuracy of dose delivery. It can be marked by a system of three intersecting laser line beams in the same way as the isocenter at a linear accelerator is marked. The aim of the test is to force the lasers to the reference setup point.

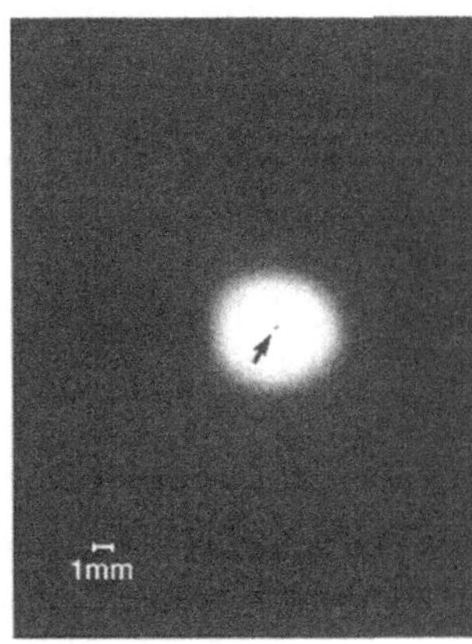

Fig. 3.10a,b. Marking the reference setup point with a laser system.
a A box with marking lines at three sides containing a film is accurately aligned to each of the laser lines. At their crossing point, a needle tip hits the film.
b After complete radiosurgical irradiation with the smallest available collimator the position of the center of dose (= reference setup point) relative to the laser lines serves as a check of correct alignment of the lasers to the reference setup point

Method. The procedure requires that the laser lines are previously adjusted exactly in horizontal and vertical directions. Proper horizontal adjustment can be easily checked by markings at the opposite wall and additionally by the agreement between "left" and "right" laser. Verification of correct vertical direction of the ceiling laser may be more difficult. It should be checked by a plumb line.

A small box made of plastic is necessary. Inside the box a radiographic film is placed. Cross markings on the surface of the box as shown in Fig. 3.10 (left) are very useful. A needle is inserted through a small drill hole at the cross center. The box is adjusted to the laser lines in such a way that the intersection point of the lasers coincides with the cross center as shown in Fig. 3.10. Next, a complete radiosurgical irradiation using a very narrow beam (4-5 mm) is applied to the box containing the film. During irradiation the tip of the needle strikes the film. The hole of the needle tip at the film and thus the laser adjustment can then be compared with the center of the radiosurgical dose distribution on the film (Fig. 3.10, right). The test has to be performed on two orthogonal planes.

Result. In principle, this method will always force the lasers to the center of the actual dose distribution regardless of how well gantry and couch rotation axes are aligned. Therefore, the method is not appropriate to check this. However, given that calculated and measured dose distribution agree well, this test provides a very accurate method to mark the reference center point.

The deviation between the needle hole and the center of the radiosurgical dose distribution should not exceed 0.2 mm.

Type. A, B

Test 23: Marking the Reference Setup Point with Floorstand (Joint Center Test)

Aim. How should the BRW floor stand and the secondary collimator be aligned to produce the most accurate treatment? The careful positioning of these two apparatus establishes a "reference setup point". This problem is complicated by such things as the fact that couch and gantry axes usually do not intersect perfectly, the gantry sag effect, and the possible wobble of the couch axis.

Method. A linac's mechanical stability, relevant to radiosurgery, should meet specifications suggested in tests 17, 18, and 19. The secondary collimator should be aligned according to test 20. A properly aligned secondary collimator will have its central axis coincident with the line connecting the radiation source (beam spot) to the axis of gantry rotation at the position nearest the axis of couch rotation. Because of gantry "sag", this can only be achieved at one gantry position. The logical choice is the intermediate or lateral gantry position.

The BRW floor stand should be set up such that the BRW (0,0,0) is on the axis of couch rotation at the height of the gantry axis of rotation. A desirable but not critical additional requirement is that the floor stand should have it's axes of travel parallel or perpendicular to the axes of couch and gantry rotation. The alignment procedure begins by setting the target simulator with a pointer attached (as opposed to the steel ball) to (0,0,0) using the BRW phantom base. (The accuracy of the phantom base (0,0,0) is confirmed by the rotational axes of the BRW arc system.) The target simulator, when attached to the floor stand, indicates the origin of the BRW system. Next, the collimator pointer is set to indicate the height of the gantry axis. This is done by adjusting the pointer length so that at gantry positions of 3 and 9 o'clock the same lateral position is indicated by the collimator pointer. Return the gantry to 12 o'clock. The collimator pointer now shows gantry axis height. Adjust the height of the floor stand on its base so that the target simulator pointer is at the height of the gantry. Begin rotating the couch and concentrate on the movement of the target simulator pointer as a function of couch rotation. Adjust the floor stand on its base until the target simulator pointer remains fixed under couch rotation. This is the reference center point. Alignment of both the secondary collimator and the floor stand is greatly facilitated by making use of the vernier motions of the floor stand to quantify the position of one pointer with respect to the other.

Result. The accuracy of this setup procedure can be measured using the procedure outlined in the general test 25, method 2 using (0, 0, 0) as the target simulator coordinates. An accuracy as stated in the Result section of Test 25 should be achieved.

Type. A, B

Test 24: Target Positioning Using Double Portal X-Ray Film

Aim. The spatial accuracy and precision of the target positioning procedure have to be assessed.

Method. This suggested method [50] again depends on accurate beam alignment to the reference setup point. After positioning the target, four plastic plates with radiographic markers are mounted on

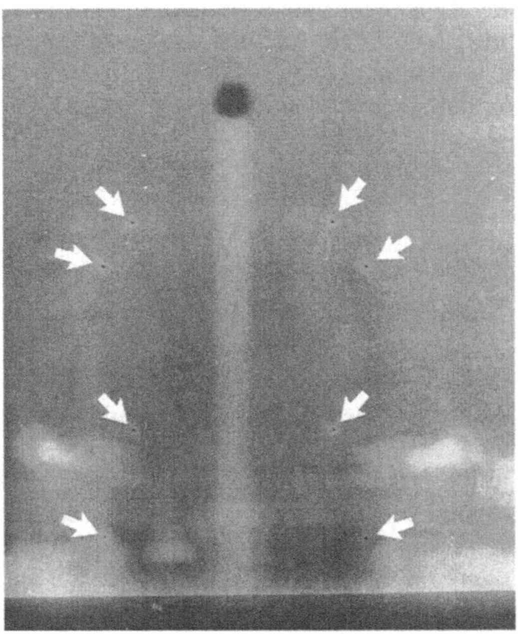

Fig. 3.11. Positioning verification with fiducial markers. The fiducials (indicated by *arrows*) and the radiosurgical beam center can be seen on the exposure

the anterior, posterior, right, and left surface of the head ring. These may be the same plates as used for angiographic localization, but with radiographic lead markers large enough to be detected on a port film. The same circular collimator which will be used for treatment is used. Anterior/posterior and right/left lateral double exposure portal films are taken. Double exposure means a first exposure with a field size sufficiently large to cover the fiducial markers and a second with the collimator inserted. This produces an image of the fiducials and radiosurgical beam center as shown in Fig. 3.11. The center of the collimator is marked. Then, the stereotactic coordinates of this center can be calculated using the angiographic localization method. This indicates the coordinates of the reference setup point in stereotactic space, which should match with the coordinates of the target point.

Result. The same key parameters as defined for target localization (see Sect. 2.1.2) should be evaluated. The method is almost as accurate as angiographic localization with typical results reported for $\bar{r}_i$ and s_r being 0.6 mm and 0.3 mm. The method thus confirms the correct patient setup prior to treatment. The resulting films may also serve to document the actual patient positioning for the patient's record file.

Type. A, C

Test 25: The General "Joint Center" Test for Target Positioning Using the Target Point Simulator

Aim. The same general test procedure can be used to address two similar aims: (1) To provide a verification that the treatment equipment is set up correctly to the target coordinates for a particular patient and that the entire system is properly aligned for that treatment and (2) To provide a general quality assurance test (not associated with a patient treatment) which can quantitatively measure the ability of the treatment equipment to center a dose distribution around an arbitrary target.

Method 1: Test Preceding Patient Treatment. This particular test applies to the linac approach using BRW equipment with a pedestal mount (BRW floor stand). This test verifies that the treatment equipment is set up correctly to the target coordinates and that the entire system is properly aligned. Alignment and setup are verified by simulating the patient's lesion with a small steel sphere positioned at the coordinates of the patient's target and making radiographic exposures with the gantry and couch in eight representative positions. Once a target center has been identified on a particular CT slice or on a particular pair of orthogonal planar angiograms, two people independently calculate the coordinates of that target. If there

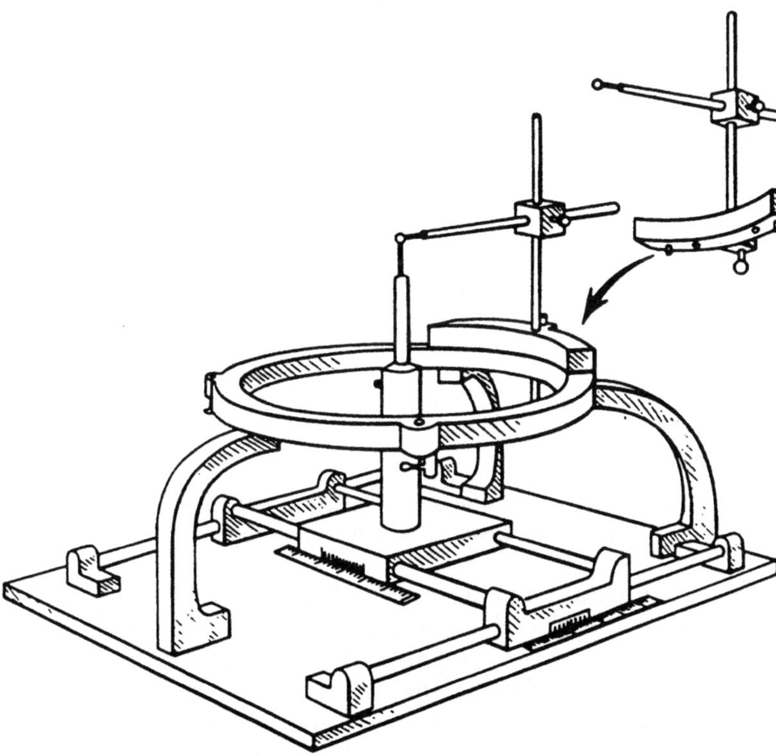

Fig. 3.12. Test of correct target positioning

is agreement, they proceed to the next step. After the collimator and BRW floor stand are fixed in place, one person sets the target coordinates on the floor stand while the other person sets the same coordinates on the BRW phantom base. The phantom base simulates the target position relative to the head ring. Next, the target simulator is attached to the phantom base (Fig. 3.12).

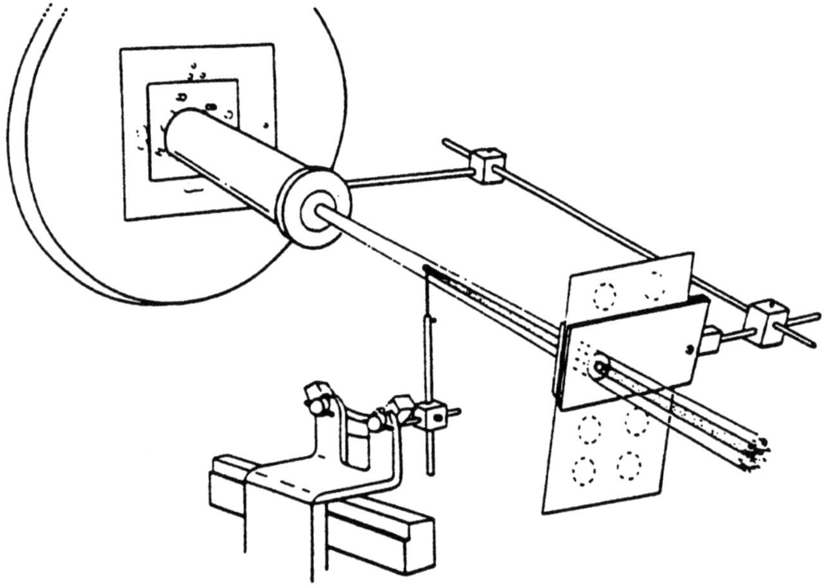

Fig. 3.13. Film holder attached to the collimator. The film is exposed with the linac X - ray beam at eight different combinations of turntable and gantry positions

The target simulator contains a section of head ring which can be attached on top of the phantom base ring. The steel ball on the simulator is then precisely aligned, using a magnifying lens as an aid, on the tip of the pointer of the phantom base. (Note that the vertical coordinate of the target must be adjusted to account for the radius of the steel ball and for the thickness of the head ring.) Next, the target simulator is carefully removed from the phantom base and attached to the BRW floor stand. The steel ball represents the patient's target and is precisely at the isocenter if no errors have been made. A film holder is now attached to the collimator (Fig. 3.13) and a film is exposed with the linac X-ray beam at eight different combinations of turntable and gantry positions. This film verifies the alignment of the entire system over the range of possible gantry and turntable movements. If the steel ball is centered within acceptable limits on all eight exposures, it can be concluded that everything is set up and aligned correctly for treatment of a target at these specified coordinates. This verification procedure will detect subtle problems with the linac (e.g. displacement of the radiation source) as well as misalignment of the supplemental collimator or BRW floor stand, and mistakes such as incorrectly setting the intended target coordinates. This verification procedure checks all positioning aspects of the treatment subsequent to determination of the target coordinates. It cannot detect an error in the target coordinates regardless of the cause of the error. Nor can this test detect an error caused by slippage of the head ring on the patient's head.

Result. When the eight exposure film is used to verify a patient setup, ball images should appear displaced by no more than approximately 1 mm (Fig. 3.14). A millimeter scaled overlay is a convenient way to measure displacement.

Method 2: General Quantitative Test of Positional Accuracy of Treatment Delivery. The procedure is exactly the same as that described above (1) except that an arbitrary set of target coordinates are chosen. This time the displacement of the ball image from the center of the radiation field is measured for each of the eight exposures. This data can be used to determine the displacement of the radiation distribution, created by the eight fields, from the target sphere and hence the geometric accuracy of the treatment delivery. This fairly messy process is described in reference [33]. However, experience has shown that using film errors only("raw" data if you will) is just about as useful in assessing accuracy as going through the more elaborate process. An individual exposure representing a particular combination of gantry and couch angles shows only a component of the actual error that exists between the center of the target and the center of the dose distribution. But this is compensated for somewhat by the fact that these errors are measured on the film (do not demagnify film error to isocenter) and are magnified by approximately 1.25. Thus the average of the eight ball-radiation field displacements on film ("raw" data) is a very good measure of the displacement of the resulting radiation distribution from the target center.

Result. Let the average displacement in a single test (average of eight exposure film) be r_i. Then the typical results after conducting many tests at different target positions for a well aligned system is $\bar{r}_i \pm s_r = 0.6$ mm $\pm$ 0.2 mm where s_r is the standard deviation of $\bar{r}_i$.

Type. A, B, C

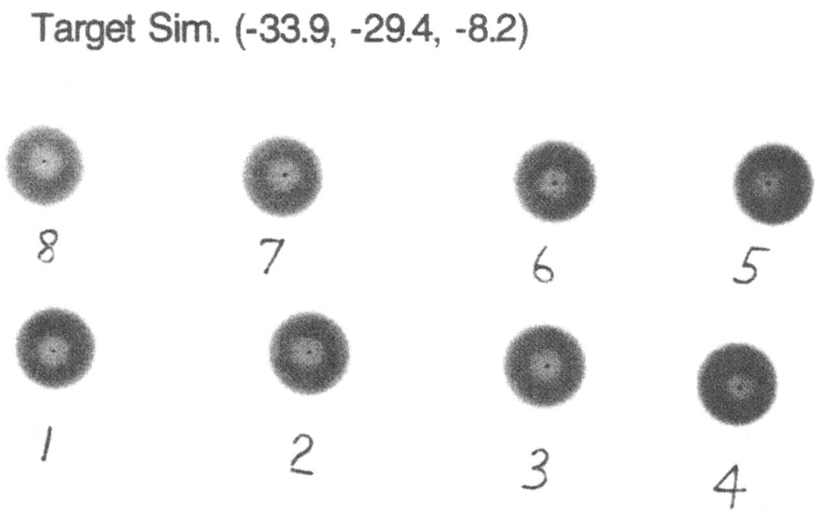

Fig. 3.14. Result from the "Joint Center Test". A film is exposed with the linac X-ray beam at eight different combinations of turntable and gantry positions. The small dots which indicate the center of the test target (steel ball) were added to help in measuring displacement errors. (Adapted from Lutz et al. [33])

Test 26: Supplementary Test for Target Positioning

Aim. Physicians feel that they have no visual way to verify, at the time of treatment, that the isocenter is positioned within their patient's target. The aim of this test is to avoid mistreatment due to incorrect localization or patient setup for treatment by a quick plausibility check.

Method. A less quantitative, supplementary test is described which checks for errors in the localization process and/or treatment setup. Using orthogonal angiographic films, CT slices or MRI slices, the distances from the lesion center to the head surface are measured along the three principal axes of the patient. Fig. 3.15 illustrates the process for CT. Measurements are made (at the CT console) from two CT slices: one through the center of the lesion (or nearly so) and one through the top of the head. The difference in table position between these two slices is the "lesion center to superior" distance.

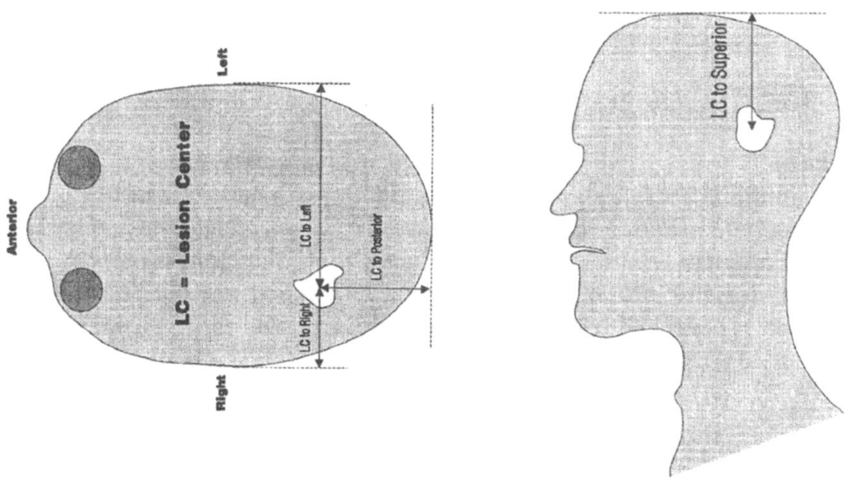

Fig. 3.15. Supplementary positioning test. Two CT slices are needed: an axial slice through the center of the lesion (*left*) and a sagittal slice through the top of the head (*right*). The distances from the lesion center to the head surface, along the three principal axes, are measured. This information then serves to check the positioning of the lesion using the lateral and ceiling lasers and a couple of rulers

These measurements are made as part of the localization process and carefully recorded. At the time of treatment and after the patient has been secured in the treatment position, these same three distances can be measured reasonably accurately on the patient using the lateral and ceiling lasers and a couple of rulers fastened together to form an "L" shape or right angle square. These distance measurements made just before treatment should agree reasonably well with the same distances calculated during the localization process.

Result. The distances from lesion center to lateral, posterior, and superior surfaces measured in these two ways - (1) from CT, MRI, or angiographic images and (2) laser lines to head surface - should agree within ± 6 mm of each other. Successful completion of this test, even with its limited accuracy, can be very reassuring.

Type. C

3.3.2.3 Method Independent Treatment Delivery Tests

Test 27: Radiographic Assessment of Beam Alignment

Aim. The calculations made with the treatment planning program are based on perfect beam alignment. A comparison between the calculated and measured dose distribution is, therefore, always a practical radiological test of the quality of the beam alignment. The comparison in particular must confirm that calculated and measured dose profiles match. Furthermore, the profile maximum must match the reference setup point.

Method. The test is based on profile measurements by film dosimetry. The exposures are made with films placed at the center of an appropriate (spherical) phantom. To allow measurements along the three principal axes, irradiations should be performed at two film plane orientations. The measured dose profiles are compared with the calculated distributions.

Two methods can be used to obtain a quantitative measure of the quality of the alignment: (1) the distance over which the dose falls from 90% to 10% is compared, and (2) the profile width (at selected isodose levels) is compared.

Result. An example of results obtained with this method is shown in Fig. 3.16. The difference in the half width of a profile should not be larger than 1 mm. Differences of 0.5 mm or less are achievable and are recommended.

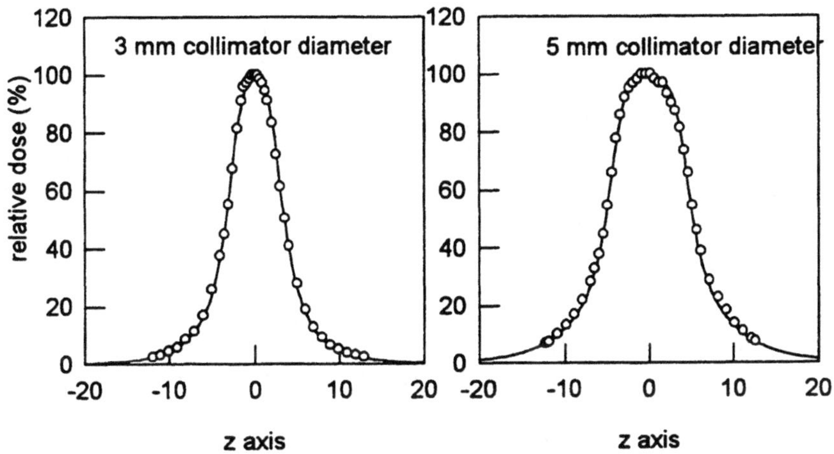

Fig. 3.16. Comparison between profiles of radiosurgical dose distributions that are measured (*dots*) and calculated (*solid line*). Six arcs of 140^0 were used in these examples. Each profile was obtained from a frontal plane along the z direction. The agreement is an indication that the actual alignment of the linac components is sufficient to provide a dose distribution as expected from the treatment planning for perfect alignment

Type. A, B

Test 28: Complete System Integration

Aim. It is important to test the entire radiosurgery system, localization through treatment, for geometric accuracy and precision. A

comprehensive method, referred to as unknown target test, provides the required information. A second method is the ring target test.
Method 1. Unknown Target Test" (Joint Center Test). To obtain quantitative results the method involves the use of test targets (steel or lead balls) of unknown position placed inside a head phantom. These unknown targets are then localized by CT or planar radiographs, under the same conditions and with the same equipment used for patients. (If MRI is used for localization, alternate target material must be used.) The head phantom is next "treated" to the target coordinates found from the localization procedure. A series of fixed fields are utilized (preferably with a small collimator) which represent entrance points spread over the upper hemisphere of the skull. A port film exposure is made for each beam. At least eight beams per target are recommended. The displacement of the image of the test target ball from the center of the field is measured for each beam. From this information the geometric error between the target point and the center of the dose distribution can be calculated (see Result in Test 25).

Result. If this test is repeated 6 to 10 times a very good understanding of the system's ideal geometric accuracy and precision can be obtained. These results help determine the necessary margin between the prescription isodose surface and the clinical target boundary that assumes full target coverage for a particular confidence level. A summary of the geometric error obtained in the "treatment" of many unknown targets at three institutions is given below in Table 3.2.

Type: A, B

Table 3.2. Results for the key parameter $\bar{r}_i$ and s_r for the "unknown target test"

Localization technique	Number of targets	$\bar{r}_i$	s_r
CT	14	1.3 mm	0.6 mm
Plain film (angiography)	5	0.5 mm	0.2 mm

Method 2. Ring Target Test. A radioactive wire and film dosimetry is needed. As an example of a radioactive wire, a Tungsten wire can be made radioactive by neutron activation. The beta-ray active Tungsten isotope W-185 with a half-life of 75.1 days is generated.

In order to measure the degree of matching between target volume and dose distribution in two dimensions, a thin radioactive wire provides a two-dimensional target structure. The purpose of this wire is twofold: (1) it is radiographically visible and hence it can be localized by CT (2) its radioactivity can be used to obtain an image of its position at a verification film. The wire is shaped into a ring of 2 -3 cm in diameter and placed between the two parts of a phantom thus representing a hidden target. The location of this target structure is performed by angiographic (Fig. 3.17a) or CT localization. If a therapy verification film is additionally placed between the two parts of the phantom during "treatment", the location of the ring can be analyzed due to the blackening by autoradiography of the ring in relation to the dose distribution (Fig. 3.17b).

Remarks. This is a complex test method which is not appropriate as a routine test. The aim of it is to test the match of the prescriptive isodose contour to a target surface rather than to check the location of the target point. The aim of this test, however, can also be accomplished by other techniques, for example, by exchanging the radioactive ring for a set of radiopaque needle tips which hit the film at points representative of a target surface.

Result. An example of applying this test is shown in Fig. 3.18. In this case, a matching between the 50% isodose line and the target ring was aimed for. The Figure shows two perpendicular dose profiles through the transverse dose distribution, the A/P and the R/L profile. The overall accuracy found in this example, that is the difference between the target and the 50% isodose, varied from 0.0 to 0.8 mm.

Type. B

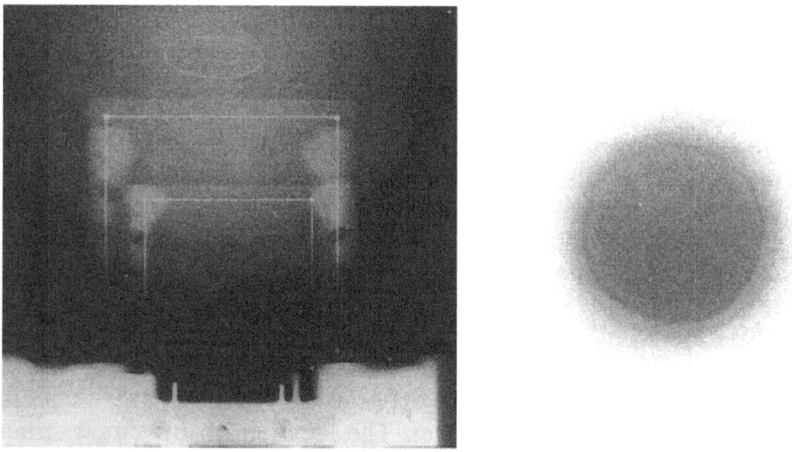

Fig. 3.17a,b. Ring target test. **a** A radioactive Tungsten ring that is detectable during X - ray or CT localization within a solid phantom is used. A film together with the ring is placed between the phantom. **b** After radiosurgical irradiation, the position of the autoradiographic pattern of the ring can be compared with the localization of the dose distribution

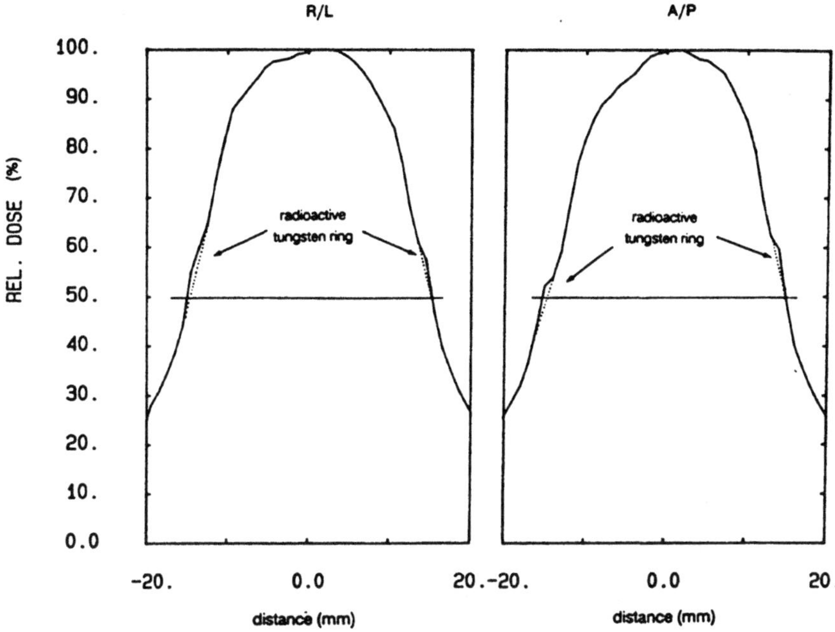

Fig. 3.18. Results from a ring target test. The 50% isodose line was aimed to match with the ring. Evaluation from profiles through the density distribution in Fig. 3.17b exhibits a deviation between the 50% isodose level and the target

4 List of Tests

4.1 Stereotactic Localization and Immobilization

Test 1: Single Point Accuracy and Precision in Angiography

Test 2: Single Point Accuracy and Precision in CT Localization

Test 3: Single Point Accuracy and Precision in MRI Localization

Test 4: "Known" Target Test During Localization with Patient

Test 5: Head Ring Movement Test

4.2 Dosimetry and Treatment Planning

Test 6: Detector Size Appropriateness

Test 7: Accuracy of Detector Positioning

Test 8: Single Beam Model for Treatment Planning

Test 9: Treatment Planning Verification

4.3 Treatment Delivery

4.3.1 Tests for Gamma Knife (Model B) Based Radiosurgery

Test 10: Air Kerma Rates in the Treatment Room

Test 11: Wipe Test

Test 12: Geometrical Accuracy Related to the Treatment Unit

Test 13: Dose Delivery and Geometrical Dose Distribution Related to the Treatment Unit

Test 14: Absolute Dose Calibration

Test 15: Treatment Position of Collimator Helmet

Test 16: Safety Features in the Gamma Knife

4.3.2 Tests for Linac-Based Radiosurgery

Test 17: Wobble of Gantry or Couch Rotation Movement

Test 18: Alignment of Gantry and Couch Rotation Axes

Test 19: Measuring the Effect of Gantry "Sag" on the Beam Axis

Test 20: Secondary Collimator Alignment

Test 21: Beam Spot

Test 22: Marking the Reference Setup Point with a Laser System (Heidelberg Test)

Test 23: Marking the Reference Setup Point with Floorstand (Joint Center Test)

Test 24: Target Positioning Using Double Portal X-Ray Film

Test 25: The General "Joint Center" Test for Target Positioning Using the Target Point Simulator

Test 26: Supplementary Test for Target Positioning

4.3.3 Method Independent Treatment Delivery Tests

Test 27: Radiographic Assessment of Beam Alignment

Test 28: Complete System Integration

Appendix A: Accuracy Requirements for Dosimetry from Tumor Control Probability Data

A meaningful statement on requirements and tolerance levels of absorbed dose can be given only on the basis of clinical observations. Following a concept given by Brahme [11], some considerations, however, can be discussed independently of the latter. These considerations refer to the detailed analysis of how variations of uncertainties in the dose absorbed in the tumor influences the tumor control probability (TCP).

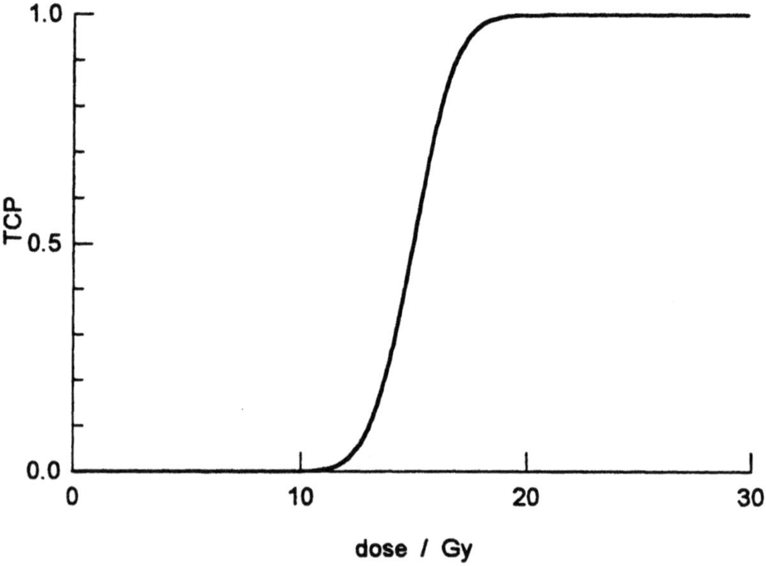

Fig. A1. Sigmoidal shape of a dose response curve to control a tumor. In this example, a dose of 15 Gy was assumed to be able to be effective in 50% of all cases

Different models of the shape of the dose response curve all result in sigmoidal curves as schematically shown in Fig. A1. The most important quantity of such a dose response relation with regard to the requirements of tolerance levels on absorbed dose is the normalized dose response gradient γ defined by:

$$\gamma = \frac{\Delta(TPC)}{\Delta D/D} = D \cdot \frac{\Delta(TPC)}{\Delta D}$$

This parameter is a dimensionless number which describes how large a change in tumor control probability is to be expected for a given relative increase in absorbed dose. For example, a dose increase of 1% will result in an increase in tumor control probability of γ%.

With this parameter, requirements of tolerance levels on absorbed dose can be discussed in relation to dose response curves more quantitatively. In particular, it turns out that a given tolerance, i.e., a possible variation in the mean dose in the target volume will lead to a certain loss of control of the treatment outcome which is higher for higher values of γ. Brahme et al. [11] concluded, that only for tumors where γ is smaller than 2, will a relative standard deviation of the mean dose of more than 5% have less than 10% of the relative standard deviation in the dose response.

Appendix B: Suggestion for Testing a Treatment Planning System

This suggested test [30] contains a list of issues that should be carefully checked before a treatment planning system (TPS) is clinically used. For this purpose, the outcome of the TPS has to be quantitatively compared with measured data. It is recommended to perform such tests with a selected set of radiosurgical beams, for example for three collimators with apertures of 5 mm, 10 mm, and 20 mm in diameter. For each single test a tolerance level for the uncertainty of the test parameter is recommended. The TPS should be clinically used only if all test results are within the given tolerance levels.

B.1 Test of the Single Field Model

Test B1

This test is designed to verify depth dose calculations for a single stationary beam. A comparison is made at the central axis between calculated and measured depths of the 80%, 70%, and 50% isodose line in a water phantom. It should be possible to calculate the measured depths with an uncertainty equal or less than 0.5 mm.

Test B2

This test is designed to verify off-axis calculations for a single stationary beam. A comparison is made at isocenter distance between calculated and measured widths of the 50%, 10%, and 5% isodose line in a water phantom. It should be possible to calculate the measured width with an uncertainty equal or less than 0.3 mm.

B.2 Test of Angle Interpolation for One Complete Rotational Irradiation

Test B3

This test is designed to verify continuously moving beam calculations. The angle increment used to simulate the moving gantry must be selected so that the calculated isodose contours match with measured isodose in a phantom. (i.e., that isodose lines will not be wavy). For a rotational irradiation, it should be possible to compute circular isodose contours: above the 50% level, deviations from a circle should be less than 0.3 mm; between the 50% and the 20% level, deviations should be less than 5 mm.

B.3 Test of Calculated Dose Profiles Obtained for Radiosurgical Dose Distributions

Test B4

This test is designed to verify the calculation of radiosurgical dose distributions. A target point is placed successively at three distances (0 cm, 2 cm, and 6 cm) from the center of a spherical phantom. It should be possible to calculate measured profile widths through the target point along the three principal axes equal or better than 0.5 mm at the 80% level, equal or better than 1.0 mm at the 50% level, and equal or better than 10.0 mm at the 20% level.

B.4 Test of Monitor Units

Test B5

This test is designed to verify calculations of the monitor units provided by the treatment system. A comparison is made between the prescribed dose to a target point in a phantom and the measured absolute dose at the target point. It should be possible to compute the monitor units required for a given dose to the target within 3%.

Appendix C: Correction of Spatial Distortion in Magnetic Resonance Imaging for Stereotactic Operation/Treatment Planning in the Brain

Aim

Depending on the individual MR system, inhomogeneities and nonlinearities induced by eddy currents during the pulse sequence can distort the images and produce spurious displacements of the stereotactic coordinates in both the x-y plane and the z axis. Such spatial distortions in magnetic resonance imaging should be measured and corrected before stereotactic localization. Errors in position can be assessed by means of two phantoms placed within the stereotactic guidance system: a two-dimensional (2D) phantom displaying "pincushion" distortion in the image, and a three-dimensional (3D) phantom displaying displacement, warp and tilt of the image plane itself. The pincushion distortion can be "corrected" by calculations based on modelling the distortion as a fourth order two-dimensional polynomial.

Method

Accurate planning of stereotactic neurosurgery [46], interstitial radiosurgery, or radiotherapy [47] requires precise spatial information. By virtue of good soft tissue contrast and multiplanar tomographic format, magnetic resonance is a logical choice providing for display of the pertinent anatomy and pathology. Accurate spatial representation demands uniform main magnetic fields and linear orthogonal field gradients. Inhomogeneity of the main magnetic field and non-linearity of the gradients produce image distortion [37]. A significant source of these geometric artifacts are eddy currents produced during the imaging sequence. Correction of these distortions is usually unnecessary for clinical diagnosis but is important for stereotaxy and in planning interstitial or external radiation therapy. A method is described to assess and correct geometrical distortions of MR images using phantom measurements.

Method 1: Two-Dimensional Phantom Measurement. At first, the two-dimensional (2D) - phantom, is used to measure the geometrical distortions within the imaging plane (Fig. C.1).

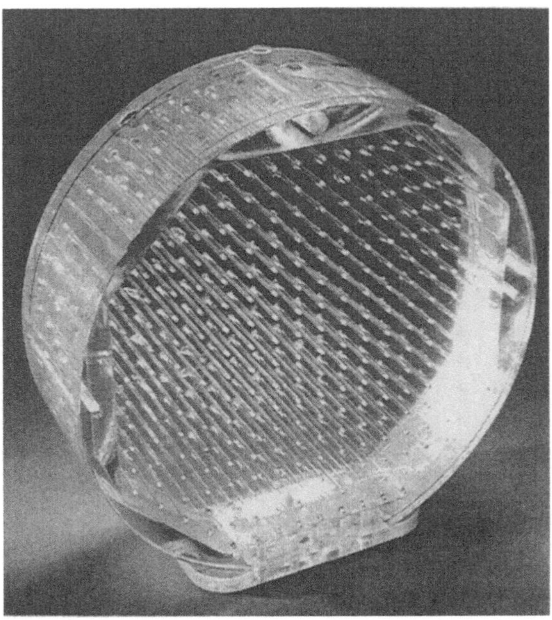

Fig. C1. The 2D - phantom used to measure geometrical distortions within the imaging plane

It consists of a water-filled cylinder 17 cm in radius and 10 cm in depth containing a rectangular grid of plastic rods spaced 2 cm apart and oriented in the z direction (i.e., perpendicular to the imaging plane).

Since the exact positions (x,y) of these rods are known a priori, their positions (u,v) in the image of the 2D - phantom reflect the geometric distortion in the imaging plane and may be used to calculate the coordinate transformation that mathematically describes the distortion process. The measuring sequence parameters include a flip angle of 15^0, a repetition time (TR) of 40 ms, an echo time (TE) of 7 ms, one acquisition, and a 256 x 256 image matrix with a 64 partition slab in the axial direction. The minimal field of vision is restricted to 260 mm in order to image the reference points of the stereotactic localization system; slab thickness is 64 mm for the 64 partitions, resulting in an effective voxel size of 1 mm^3.

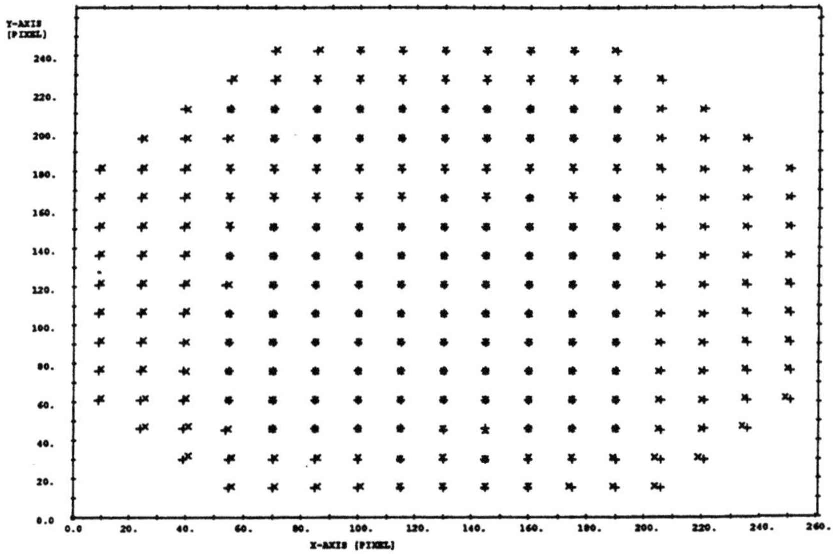

Fig. C2. True pin positions (+) and image (distorted) positions (x) of the 2D phantom of Fig. C1

The x-y plane distortion of axial MR images of the 2D-phantom can be "corrected" (reducing displacements to the size of a pixel) by calculations based on modeling the distortion as a fourth order two-dimensional polynomial:

$$u_k = \sum_{i=0}^{N}\sum_{j=0}^{N} U_{ij} \cdot x_{ki} \cdot y_{ki}$$

$$v_k = \sum_{i=0}^{N}\sum_{j=0}^{N} V_{ij} \cdot x_{ki} \cdot y_{ki} \qquad \text{with } k = 1,2,3,\ldots M$$

where (x,y) are the true pin positions (+ in Fig. C2), (u,v) are the distorted positions measured in the image (x in Fig. C2), N (= 4) is the order of the polynomial, and M (= 249) is the total number of pin positions of the 2D-phantom. Note that the origin of the u,v coordinate system corresponds to the origin of the x,y coordinate system and that it lies in the center of the image, that is, the area free of distortion.

Equation 1 and 2 applied to all M pin positions of the 2D - phantom form a system of simultaneous linear equations from which the coefficients U_{ij} and V_{ij} can be calculated and the distortion corrected. Thereby the selection of $N = 4$ is a compromise between computation time and reduction in distortion.

A more quantitative, direct measurement of distortion is provided by the distortion vector which measures the discrepancy between the true $(x,y)_k$ position of a pin in the 2D - phantom and its apparent position in the image $(u,v)_k$: $d_k = |(x,y)_k - (u,v)_k|$ (i.e., the vector distance between the true and apparent positions). Reduction in distortion is demonstrated in Fig. C3 - a comparison of the distribution of d_k in the uncorrected and corrected images. The distribution of d_k in the corrected 2D - phantom image shows, that the positional errors are reduced from about 2-3 mm (Fig. C3, solid line) to about 1 mm across the entire 2D - phantom (Fig. C3, dotted line). A further attempt to reduce this residual error does not appear meaningful since this approximates the dimensions of the image pixels (256 x 256 matrix).

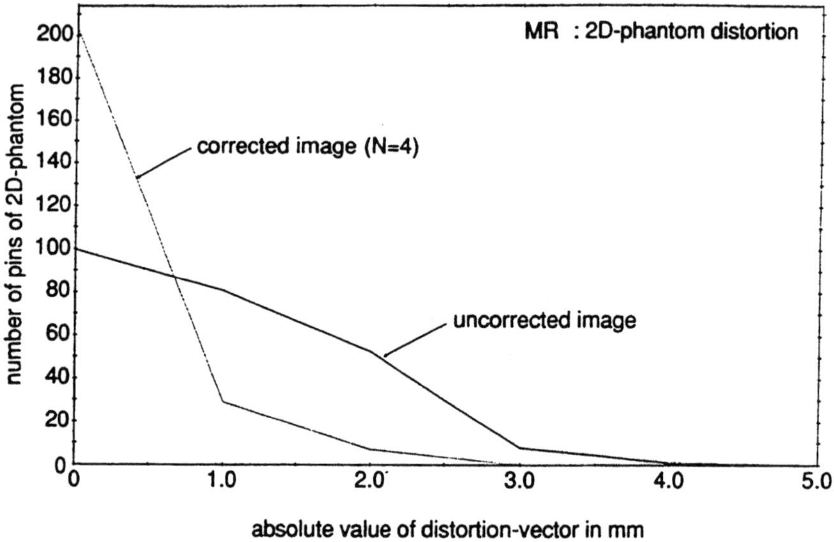

Fig. C3. MR: 2D phantom distortion. Comparison of the distribution of the length of distortion vectors in the uncorrected (*solid line*) and corrected (*dotted line*) images. The positional errors are reduced from about 2-3 mm to about 1 mm

Method 2: Three-Dimensional Phantom Measurement. After correcting for distortion in the imaging plane, the three-dimensional position of the imaging plane was assessed with the 3D - phantom (Fig. C4). This is necessary because the inhomogeneities of the gradient fields that define the imaging plane can produce deformation or tilting of the plane in space which is not apparent from the 2D - phantom measurements. For the measurements the 3D - phantom is fixed in the stereotactic guidance system and axial orientated MR images are obtained and initially corrected using the 2D polynomial.

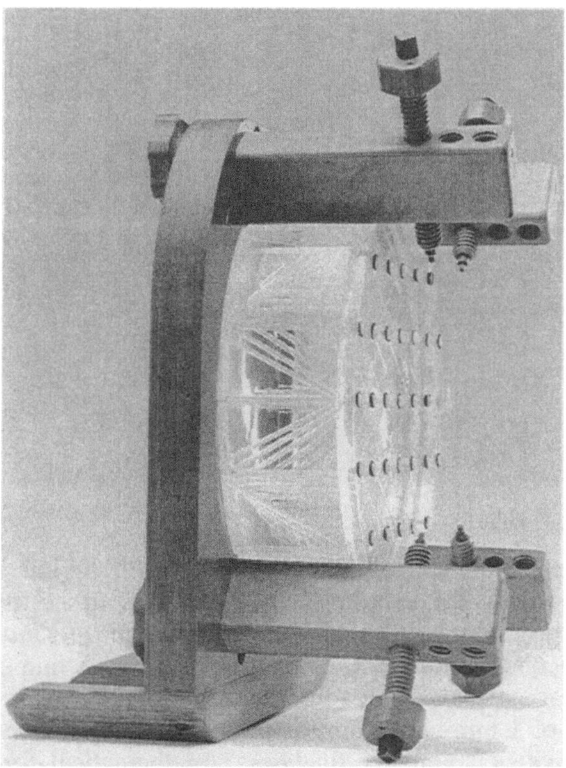

Fig. C4. The 3D-phantom for measuring geometrical distortions within the imaging plane. It contains rectangular as well as oblique water-filled boreholes

These corrected images display a regular pattern of points emanating from the water-filled boreholes (Fig. C5). Each reference point produced by a rectangular borehole is encircled by several measurement points coming from neighbouring oblique boreholes. The distance between the reference and measurement points is a direct measure of the z coordinate of the imaging plane at the reference point (Fig. C6).

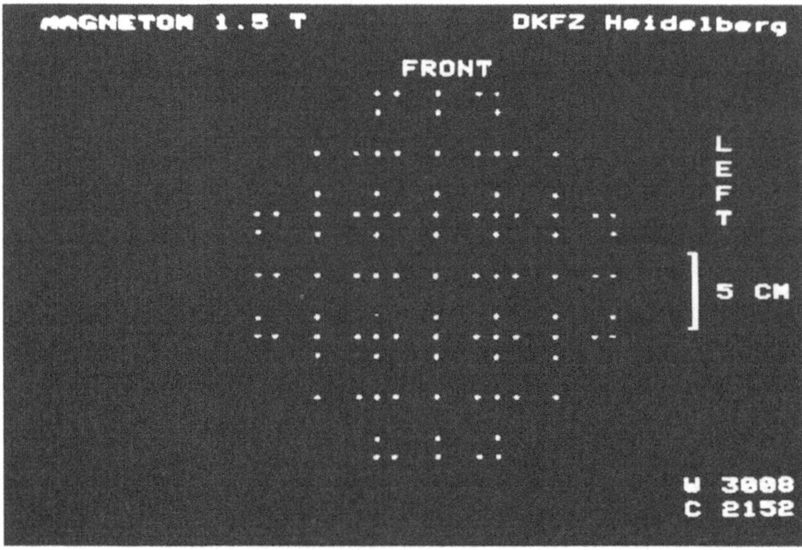

Fig. C5. Image produced by the 3D - phantom of Fig. C4

In this manner at every reference point the z coordinates of the imaging plane can be measured, the imaging plane reconstructed and its position, shape, and orientation assessed. Fig. C7 is a typical example of the initial reconstruction of the imaging plane showing a nearly horizontal imaging plane with some discrepancies in z components to the order of 1 mm which approximates again the dimensions of the image pixels (1 mm slice thickness). Mathematical correction of these discrepancies in the z direction was not pursued since properly adjusted shims avoid deformation or tilting of the imaging plane.

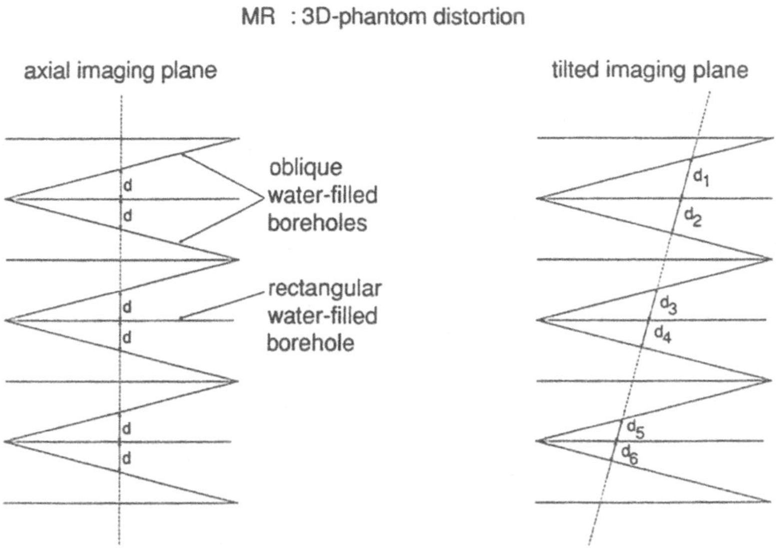

Fig. C6. MR: 3D phantom distortion. The distance *d* between the reference and measurement points is a direct measure of a possibly tilted imaging plane

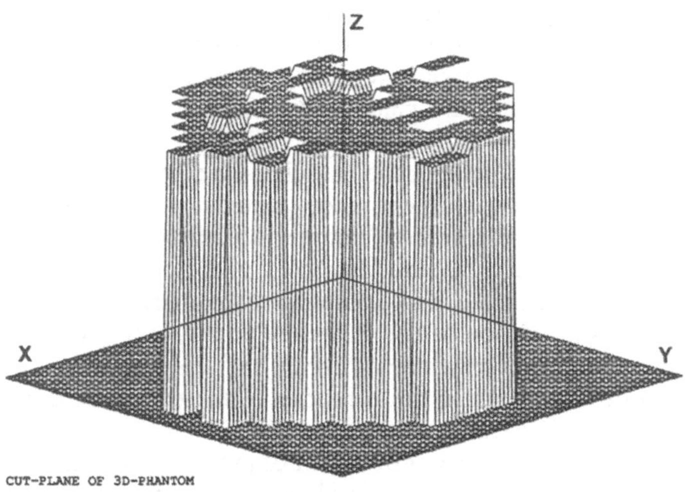

Fig. C7. A typical example of the imaging plane showing a nearly horizontal imaging plane with some discrepancies in *z* components to the order of 1 mm

Result

The precision of MRI stereotaxy and the importance of geometric corrections can be tested by assessing the stereotactic coordinates of a point source with and without correction and comparing them with those measured mechanically.

As an example, a cross plastic tube (1 mm inner diameter) filled with Gd-DTPA solution was placed in the stereotactic guidance system. The cross was aligned in the z direction and the stereotactic coordinates of its crosspoint was measured (3D-FISP sequence) from axial MR images. The stereotactic z coordinate (i.e., the distance of the image plane from the stereotactic zero plane) was determined by the distance between two reference points produced by the obliquely oriented plastic tubes of the guidance system [45]. The stereotactic x and y coordinates were measured with respect to the stereotactic zero point (midpoint of the guidance system). The manually measured exact stereotactic coordinates of the crosspoint were $P_{man}(x,y,z) = (48.5, -27.5, 78.5)$ mm compared to the image-based evaluated coordinates $P_{ima}(x,y,z) = (47.6, -27.7, 78.0)$ mm, whereas discrepancies of about 2 mm appeared mainly in the z coordinate without correction of geometrical distortion. The magnitude of the error in z coordinate increases with the z coordinate. This is related in part to the pincushion-like distortion pattern of the axial images, since z coordinates are calculated from and therefore ultimately dependent on the fidelity of the x,y components. In both the uncorrected and corrected images the calculated z component correspond to the distance between the x,y components of two reference points. Ideally, and in the corrected images, this distance increases linearly with the true z component. However, in the case of the uncorrected images the distance, due to the pincushion effect which is not negligible at the position of the reference points, increases nonlinearly - the z component measured lying on an "arc" rather than on a proper "straight line". Hence the disparity of these two measurements increases with increasing z coordinate.

As a second example, an anatomic check of the correction method (i.e., correction of anatomic images via phantom measurements) was done by assessing the stereotactic position of the middle cerebral artery and comparing these with those obtained using the same stereotactic system with CT [45,46]. This proves that after correction the locations in CT and MRI correspond to the same anatomic focus. After correction, the accuracy of the geometric information is limited only by the pixel resolution of the image (= 1 mm). For example such a correction provides for the more accurate transfer of anatomical/pathological information and target point coordinates to the isocenter of the therapy machine, which is so essential for the stereotactic radiation technique, or in the event of stereotactic guided operation planning.

References

1. AAPM Task Group 24 (1984) Physical aspects of quality assurance in radiation therapy. AAPM Report No. 13, American Institute of Physics, NewYork
2. Adler JR (1993) A novel system for frameless stereotactic radiosurgery. Acta Radiochir 122: 182
3. Andreo P, Brahme A (1986) Stopping power data for high energy photon beams. Phys Med Biol 31: 839-858
4. Barcia-Solario JL, Hernandez G, Broseta J, Gonzalez-Darder J, Ciudad J (1982) Radiosurgical treatment of carotid-cavernaous fistula. Appl Neurophysiol 45: 520-522
5. Bardash M, Amols HI, Kohn S, Martel MK; Wuu CS (1992) Rapid dose calculations for stereotactic radiosurgery. Med Phys 19: 965-970
6. Betti OO, Derechinski VE (1984) Hyperselective encephalic irradiation with linear accelerator. Acta Neurochir Suppl 33: 385-390
7. Bianciardi L, D'Angelo L, Gentile FP, Benassi M, Guerra AS (1989) Dosimetry of small X - ray radiation fields. In : Clinical dosimetry. IAEA-SM-298/42 Vienna, pp 355-363
8. Bjarngard BE, Tsai J-S, Rice RK (1990) Doses on the central axis of a narrow 6-MV X - ray beam. Med Phys 17: 794-799
9. Bosch DA (1986) Stereotactic techniques in clinical neurosurgery. Springer Verlag, Wien New York
10. Brahme A (1981) Correction of a measured distribution for the finite extention of the detector. Strahlentherapie 157: 258-259
11. Brahme A (ed) (1988) Accuracy requirements and quality assurance of external beam therapy with photons and electrons. Acta Oncol Suppl 1
12. Chiriego G, Avanzo RC, Marchetti C, Scalchi P, Pozza F (1987) Small field dosimetry. In: Proceedings of the Fifth Varian European Clinac Users Meeting, pp 27-30
13. Colombo F, Benedetti A, Pozza F, Avanzo RC, Marchetti C, Chierego G, Zanardo A (1985) External stereotactic irradiation by linear accelerator. Neurosurgery 16: 154-159
14. Dawson J, Schroeder NJ, Hoya JD (1986) Penumbral measurement in water for high energy X - rays. Med Phys 13: 1011 -1114
15. Drzymala RE (1991) Quality assurance for linac-based stereotactic radiosurgery. In: Starkschall G, Horton J (eds) Proceedings of an American college of medical physics symposium. pp 121-129
16. Engler MJ, Jones GL (1984) Small - beam calibration by 0.6 - and 0.2 cm^3 ionization chambers. Med Phys 11: 822-826
17. Ermakov I, Shtukovsky O, Vatnitsky S, Chervjakov A, Kiivinitty K (1989) Dose measurements of narrow photon beams. In: Proceedings of an Interna-

18. tional Workshop on Proton and Narrow Photon Beam Therapy. 8-10 June 1989, Oulu, Finland, pp 131-138
19. Friedman WA, Bova FJ (1989) The university of Florida radiosurgery system. Surg Neurol 32: 334-342
20. Gill SS, Thomas DGT, Warrington AP, Brada M (1991) Relocatable frame for stereotactic external beam radiotherapy. Int J Radiat Oncol Biol Phys 20: 599-603,
21. Greitz T, Lax I, Bergstrom M, Arndt J, Berggren B-M, Blomgren H, Boethius J, Lindquist M, Ribbe T, Steiner L (1986) Stereotactic radiation therapy of intracranial lesions. Methodologic Aspects. Acta Radiol 25: 81-89
22. Hariz MI, Henriksson R, Löfroth P-O, Laitinen LV, Säterborg N-E (1990) A non-invasive method for fractionated stereotactic irradiation of brain tumors with linear accelerator. Radiother Oncol 17: 57-72
23. Hartmann GH, Bauer-Kirpes B, Serago CF, Lorenz WJ (1994) Precision and accuracy of stereotactic convergent beam irradiation from a linear accelerator. Int J Radiat Oncol Biol Phys 28: 481-492
24. Hartmann GH, Schlegel W, Sturm V, Kober B, Pastyr O, Lorenz WJ (1985) Cerebral radiation surgery using moving field irradiation at a linear accelerator facility. Int J Radiat Oncol Biol Phys 11:1185-1192
25. Heilbrun MP, Roberts TS (1985) CT stereotactic guidance systems. In: Wilkins RH, Rengachary SS (eds) Neurosurgery. McGraw-Hill, New York, pp 2481-2489
26. ICRU Report 23 (1973) Measurement of absorbed dose in a phantom irradiated by a single beam of X or gamma rays. International Commission on Radiation Units and Measurements, Washington D. C.
27. ICRU Report 24 (1976) Determination of absorbed dose in a patient irradiated by beams of X or gamma rays in radiotherapy. International Commission on Radiation Units and Measurements, Washington D. C.
28. ICRU Report 50 (1984) Prescribing, recording, and reporting photon beam therapy. International Commission on Radiation Units and Measurements, Washington D. C.
29. Kjellberg RN (1979) Stereotactic Bragg peak proton radiosurgery method. In: Szikla G (ed) Stereotactic Cerebral Irradiation, Elsevier, Amsterdam, pp 93-100
30. Kosunen A, Jarvinen H, Vatnitsky S, Ermakov I, Chervakov A, Kulmala J, Pitkanen M, Vayrynen T, Vaananen A (1991) Intercomparison of radiotherapy treatment planning systems using calculated and measured dose distributions for external photon and electron beams. STUK-A98, Helsinki, Finland
31. Leksell L (1971) Stereotaxis and radiolurgery. An operativc system. Thomas, Springfield
32. Lutz W, Maleki N, Bjärngard BE (1988) Evaluation of a beam spot camera for megavoltage X - rays. Med Phys 15: 614-617
33. Lutz W, Winston KR, Maleki N (1988) A system for stereotactic radiosurgery with a linear accelerator. Int. J. Radiat. Oncol Biol. Phys. 14: 373-381

34. Luxton G, Jozsef G, Astrahan MA (1991) Algorithm for dosimetry of multiarc linear-accelerator stereotactic radiosurgery. Med. Phys 18: 1211-1221
35. Lyman JT, Phillips MH, Frankel KA, Fabrikant JI (1989) Stereotactic frame for neuroradiology and charged particle Bragg peak radiosurgery of intracranial disorders. Int J Radiat Oncol Biol Phys 16:1615-1621
36. McGinley PH, Butker EK, Crocker IR, Landry JC (1990) A patient rotator for stereotactic radiosurgery. Phys Med Biol 35: 649-657
37. O'Donnell M, Edelstein WA (1985) NMR imaging in the presence of magnetic field inhomogeneities and gradient field nonlinearities. Med Phys 12: 20-26
38. Olivier A, Peters TM, Bertrand G (1985) Stereotaxic systems and apparatus for use with MRI, CT and DSA. Appl Neurophysiol 48: 94-96
39. Philips MH (ed) (1993) Physical aspects of stereotactic radiosurgery. Plenum, New York
40. Pike B, Podgorsak E, Peters TM, Pla C (1987) Dose distributions in dynamic stereotactic radiosurgery. Med Phys 14: 780-789
41. Podgorsak E, Pike GB, Oliver A, Pla M, Souhami L (1989) Radiosurgery with high energy photon beams: a comparison among techniques. Int J Radiat Oncol Biol Phys 16: 857-865
42. Podgorsak EB (1992) Physics for radiosurgery with linear accelerators. Stereotact Radiosurg 1: 9-34
43. Podgorsak EB, Olivier A, Pla M, Lefebvre PY, Hazel J (1988) Dynamic stereotactic radiosurgery. Int J Radiat Oncol Biol Phys 14:115-126
44. Rice J, Hansen JL, Svensson GK, Siddon RL (1987) Measurements of dose distributions in small beams of 6 MV X - rays. Phys Med Biol 32: 1087-1099
45. Schad LR, Boesecke R, Schlegel W, Hartmann G, Sturm V, Strauss L, Lorenz WJ (1987) Three dimensional image correlation of CT, MR, and PET studies in radiotherapy treatment planning of brain tumours. J Comput Assist Tomogr 11: 948-954
46. Schad LR, Ehricke H-H, Wowra B, Layer G, Engenhart R, Kauczor H-U, Zabel H-J, Brix G, Lorenz WJ (1992) Correction of spatial distortion in magnetic resonance angiography for radiosurgical treatment planning of cerebral arteriovenous malformations. Magn Reson Imaging 10: 609-621
47. Schlegel W, Scharfenberg H, Doll J, Hartmann G, Sturm V, Lorenz WJ (1984) Three dimensional dose planning using tomographic data. In: Proceedings of 8th International Conference on the Use of Computers in Radiation Therapy. IEEE Computer Society, Silver Spring, pp 191-196
48. Schlegel W, Scharfenberg H, Doll J, Pastyr O, Sturm V, Netzeband G, Lorenz WJ (1982) CT-Images as the basis of operation planning in stereotactical neurosurgery. In: Proceedings of 1st International Symposium on Medical Imaging and Image Interpretation. IEEE Computer Society, Silver Spring, pp 172-177
49. Serago CF, Houdek PV, Hartmann GH, Saini DS, Serago ME, Kaydee A (1992) Tissue maximum ratios (and other parameters) of small circular 4, 6, 10, 15 and 24 MV X - ray beams for radiosurgery. Phys Med Biol 37: 1943-1956

50. Serago CF, Lewin AA, Houdek PV, Gonzales-Arias S, Hartmann GH, Abitbol AA, Schwade JG (1991) Stereotactic target point verification of an X - ray and CT localizer. Int J Radiat Oncol Biol Phys 20: 517-523
51. Shibata CH, Motta HC, Beddar AS, Higgins PD, Shin K (1991) Influence of detector size in photon profile measurement. Phys Med Biol 36: 621-631
52. Sixel KE, Podgorsak EB (1993) Buildup region of high energy X - ray beams in radiosurgery. Med Phys 20: 761-764
53. Sturm V, Pastyr O, Schegel W, Scharfenberg H, Zabel H-J, Nezeband G, Schabbert S, Berberich W (1983) Stereotactic computer tomography with a modified Riechert-Mundinger device as the basis for integrated stereotactic neuroradiological investigations. Acta Neurochir 68: 11-17
54. Svenson GK (1984) Quality assurance in radiation therapy: physical efforts. Int J Radiat Oncol Biol Phys 10 (Suppl): 23-29
55. Svensson H, Brahme A (1986) Recent advances in electron and proton dosimetry. In: Orton CJ (ed) Radiation dosimetry: Physical and Biological aspects. Plenum, New York, pp 86-102
56. Tsai J-S, Buck BA, Goran RT, Svensson K, Alexander E, Cheng C-W, Mannarino EG, Loeffler JS (1991) Quality assurance in stereotactic radiosurgery using a standard linear accelerator. Int J Radiat Oncol Biol Phys 21: 737-748
57. Turner JW, Shaw A (1974) A versatile stereotaxic system based on cylindrical coordinates and using absolute measurements. Acta Neurochir Suppl 21: 211-220
58. Vatnitsky S, Jarvinen H (1993) Application of a natural diamond detector for the measurements of relative dose distributions in radiotherapy. Phys Med Biol 38: 173-84
59. Vatnitsky SM, Khrunov VS, Fominych VI, Schuele E (1993) Diamond detector dosimetry in medical applications. Radiation Protection Dosimetry 47: 515-518
60. Winston KR, Lutz W (1988) Linear accelerator as a neurosurgical tool for stereotactic radiosurgery. Neurosurgery 22: 454-463
61. Wu A (1992) Physics and dosimetry of the Gamma Knife. Stereotact Radiosurg 1: 35-50
62. Wu A, Zwicker RD, Kalend AM, Zeng Z (1993) Comments on the dose measurements for a narrow beam in radiosurgery. Med Phys 20: 777-779

GPSR Compliance

The European Union's (EU) General Product Safety Regulation (GPSR) is a set of rules that requires consumer products to be safe and our obligations to ensure this.

If you have any concerns about our products, you can contact us on ProductSafety@springernature.com

In case Publisher is established outside the EU, the EU authorized representative is:

Springer Nature Customer Service Center GmbH
Europaplatz 3
69115 Heidelberg, Germany

Batch number: 10704701

Printed by Printforce, the Netherlands